Future-proofing Travel

Future-proofing Travel

*How to create a resilient
and sustainable industry*

Caroline Bremner

KoganPage

First published in Great Britain and the United States in 2025 by Kogan Page Limited

Kogan Page
Kogan Page Ltd, 2nd Floor, 45 Gee Street, London EC1V 3RS, United Kingdom
Kogan Page Inc, 8 W 38th Street, Suite 90, New York, NY 10018, USA
www.koganpage.com

EU Representative (GPSR)
Authorised Rep Compliance Ltd, Ground Floor, 71 Lower Baggot Street, Dublin D02 P593, Ireland
www.arccompliance.com

Kogan Page books are printed on paper from sustainable forests.

ISBNs
Hardback   978 1 3986 1919 7
Paperback   978 1 3986 1917 3
Ebook   978 1 3986 1920 3

British Library Cataloguing-in-Publication Data
A CIP record for this book is available from the British Library.

Library of Congress Control Number
2024058675

Typeset by Integra Software Services, Pondicherry
Print production managed by Jellyfish
Printed and bound by CPI Group (UK) Ltd, Croydon CR0 4YY

For Mum and Dad, Finn and Ingrid

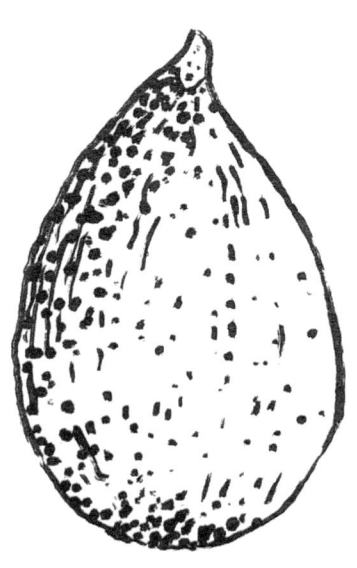

CONTENTS

LIST OF FIGURES AND TABLE

ABOUT THE AUTHOR

Caroline Bremner is a global travel thought leader with 28 years' experience in market research. Caroline advises global travel brands, destinations and consumer goods players on how to navigate the future of travel, embrace consumer trends, digitalization and sustainability. Caroline began her career in travel as a tour guide at one of the city's leading visitor attractions, Camera Obscura, after graduating from Edinburgh University with an MA (Hons) in French. Caroline also has a Postgraduate Diploma in European Marketing and Languages from Edinburgh Napier University.

As a global thought leader, Caroline is quoted regularly in the international press and is a frequent Keynote speaker on the conference circuit such as WTM London and ITB Berlin. She has written and spoken extensively about the future of travel, addressing themes like the Traveller of the Future, digital transformation and the need for sustainable tourism transformation for positive impacts for people and places.

PREFACE

The travel industry is an economic powerhouse generating trillions of dollars annually and millions of jobs, regarded as a driver of shared prosperity and sustainable development when managed correctly. It stands at a difficult crossroads after overcoming its greatest existential threat in the pandemic – the temporary shutdown of non-essential travel. Enormous resilience has been exhibited by travel businesses with public sector support, as record-breaking visitor numbers are travelling again. The return to business as usual has reawakened concerns about overtourism challenges in popular destinations, leading to counter measures such as tourist taxes, visitor quotas and a community backlash where pressures on the environment, housing, infrastructure and resources are acute.

Travel is not just a victim of its own success with places 'loved to death' but at the forefront of the climate emergency, especially small island developing states that rely on tourism, promoting their natural and cultural assets. Travel as a force for good can lift up communities, especially women and young people, regenerate fragile ecosystems and reverse biodiversity loss. Yet there is an awkward truth at its heart. The travel industry is a significant source of carbon emissions, especially transport, producing over 8 per cent globally. It bears a huge responsibility to decarbonize in line with the Sustainable Development Goals and net zero pathway to 2050.

Travel continues to be a top consumer priority where people live out their aspirations, dreams and follow their passions inspired by social media. Every generation enjoys enriching travel experiences where the middle classes in emerging markets are increasingly exploring the world. The current trajectory of future demographics, demand and supply will lead to further negative impacts on local destinations and communities, requiring the need to accelerate sustainable transformation. A paradigm shift is called for to a value-driven model. There are trailblazers leading the charge such as B Corp and Science Based Targets Initiative, as well as all those signing the Glasgow Declaration for Climate Action. Yet the industry is highly fragmented and it is extremely challenging to scale sustainable practices. This weakness can also be a strength when it comes to the role travel and tourism businesses can play in destination stewardship, value creation, diversity, equity and inclusion. Empowering grassroots action for regenerative and sustainable business practices can be accelerated at speed by digitalization, green technology and innovation.

Technology is a powerful enabler of transformation, where each exponential leap forward opens up new challenges and opportunities when technology is harnessed for good purposes. Nothing should be off the table from renewable energy, Generative AI, biometrics, web 3.0 and blockchain to next-generation mobility solutions. Digital drives personalization and Gen AI is ushering in a new era which can be used to drive awareness and engagement about impacts, with rewards and incentives for positive travel behaviours.

The future traveller will be younger, more diverse, hyper-connected, impact-conscious, driven by purpose and passion. Aligning with their values will create shared value to drive the necessary transformation that elevates the benefits and minimizes the costs to everyone across the chain.

Decoupling climate impacts from economic growth, prioritizing people and places, while doubling down on truly authentic, trusted and enriching experiences requires bold, collective action. No one can achieve this alone. Partnership, collaboration and interdisciplinary engagement are key for finding holistic solutions where travel is part of the broader systems transformation. Travel should be part of the solution, not the problem and should not work in silo.

There are signs of the costs involved for funding the transition through higher prices and eco-surcharges as costs are passed onto consumers. It is important to ensure travel does not become exclusive, which would stifle its inherent qualities of cultural exchange and understanding of different cultures and beliefs.

Future-proofing travel requires taking a long-term view while adopting an iterative approach, learning and bouncing back quickly. The travel businesses of the future will be positioned at the nexus of technology, climate, nature and people for a powerful, agile and inclusive model. Success is never guaranteed but failure is not an option when the stakes are so high. We owe it to future generations to act now. Collectively, we need to turbo-charge sustainable transformation, moving to a regenerative and net positive position. That way everyone will have the privilege of enriching travel experiences that empower positive change for travellers, communities and nature for a thriving future.

ACKNOWLEDGEMENTS

I would like to extend a huge thank you to: Susan Furber for the unwavering support; Roger Elliot for the beautiful drawings; Finn Elliot for research on permissions and feedback; Ingrid Elliot for Gen Alpha insights and support.

A huge thank you to everyone who shared their valuable insights, data and time to contribute and shape the findings of this book: Amy Jukes, Anna Leask, Ben Lyman, Chris Doyle, Chris Imbsen, Christina Beckmann, Dee Gibson, Eirik Skjærseth, Eric Ricaurte, Fiona Jeffery, Fransua Vytautas Razvadauskas, Ian Yeoman, Jake Haupert, James McDonald, Jeremy Sampson, John Sage, Jonathan Mitcham, Joshua Ryan Saha, Laura Kotyga, Marion Phillips, Nejc Jus, Rochelle Turner, Scott Wayne, Shannon Stowell, Sumeetra Ramakrishnan, Susanne Becken, Susanne Etti, Vicky Smith, Vince Shacks, Wouter Geerts and Xavier Font.

For being there, thank you to Edith and Bobby, Bremners North, Bremners South, Alex, Annabel, Pam, Adrienne, Karen, Dan, Derek and Jill, Jonathan Fisher, Prudence Lai, the Nairn girls and everyone at SoupSmiths.

LIST OF ABBREVIATIONS

5G	fifth generation wireless technologies
6G	sixth generation wireless technologies
AAM	Advanced air mobility
ACI	Airports Council International
ADA	Americans with Disabilities Act
ADR	average daily room rate
AFIR	Alternative Fuel Infrastructure Regulation
AGI	artificial general intelligence
AI	artificial intelligence
API	application programming interface
AR	augmented reality
ASA	Advertising Standards Authority
ASEAN	Association of Southeast Asian Nations
ATTA	Adventure Travel Trade Association
B2B	business to business
B2C	business to consumer
B5G	Beyond 5G
BIPOC	Black Indigenous and People of Colour
BNPL	buy now pay later
BRI	Belt and Road Initiative
BTA	Black Travel Association
CAGR	compound annual growth rate
CBD	Convention on Biodiversity Diversity
CBDC	central bank digital currency
CCUS	carbon capture, utilization and storage
CEO	Chief Executive Officer
CFTA	carbon footprint tracking apps

CLIA	Cruise Lines International Association
COP	Conference of the Parties
CORSIA	Carbon Offsetting and Reduction Scheme for International Aviation
CRM	customer relationship management
CRS	central reservations system
CSRD	Corporate Sustainability Reporting Directive
CTO	Caribbean Tourism Organization
DACCS	Direct Air Carbon Capture and Storage
DASTA	Designated Areas for Sustainable Tourism Administration
DEI	diversity, equity, inclusion
DEIA	diversity, equity, inclusion and accessibility
DMC	destination management company
DMOs	Destination Management/Marketing Organizations
DSA	Digital Services Act
EC	European Commission
ECHR	European Court of Human Rights
EMEA	Europe, Middle East and Africa
EO	earth observation
ESG	environmental, social, governance
ETIAS	European Travel Information and Authorization System
EU	European Union
EU-ETS	European Trading Scheme
EUDI	EU Digital ID Wallet
EV	electric vehicle
eVTOL	electric vertical takeoff and landing
FAA	Federal Aviation Administration
FFPs	frequent flyer programmes
FOMO	fear of missing out
FY	financial years
GARN	Global Alliance for the Rights of Nature

GBF	Global Biodiversity Framework
GBTA	Global Business Travel Association
GCC	Gulf Cooperation Council
GDPR	General Data Protection Regulation
GDS	Global Distribution Systems
GHG	greenhouse gases
GIS	geographical information system
GLSR	Green Lodging Sustainability Report
GPT	Generative Pretrained Transformer
HIA	hydrogen in aviation
HNWI	high net worth individual
IATA	International Air Transport Association
ICAO	International Civil Aviation Organization
ICEV	internal combustion engine vehicles
ICT	Internet and Communications Technology
IDB	Inter-American Development Bank
IEA	International Energy Agency
IFC	International Finance Cooperation
IGLTA	International Gay and Lesbian Travel Association
IMF	International Monetary Fund
IoT	Internet of Things
IPBES	InterGovernmental Science Policy Platform on Biodiversity and Ecosystem Services
IPCC	InterGovernmental Panel on Climate Change
IRENA	International Renewable Energy Agency
ISS	International Space Station
IUCN	International Union for Conservation of Nature
LCAF	low carbon aviation fuel
LCCs	low-cost carriers
LEED	Leadership in Energy and Environmental Design
LEO	low earth orbit

LGBTQ+	Lesbian Gay Bisexual Transgender Queer Plus
lidar	light detection and ranging
LLM	large language model
LNG	liquified natural gas
LTAG	Long-term Aspirational Goal
MaaS	mobility as a service
MICE	meetings, incentives, conventions and exhibitions
MSMEs	Micro and Small Medium-sized Enterprises
MVP	minimum viable product
NASA	National Aeronautics and Space Administration
NDC	new distribution capability
NDCs	Nationally Determined Contributions
NFT	non-fungible token
NGOs	Non-Governmental Organizations
NHSRCL	National High Speed Rail Corporation Ltd
NRP	National Rail Plan
NZE	net zero emission
NZV	net zero vehicle
OIC	Organization of Islamic Cooperation
OTA	Online Travel Agent
OUV	Outstanding Universal Value
PATA	Pacific Asia Travel Association
PMS	property management system
PV	photovoltaic
PwD	People with Disabilities
QR	quick response code
REITs	real estate investment trusts
RevPAR	revenue per available room
RMS	revenue management system
ROI	return on investment
SaaS	software as a service

SAE	Society of Automotive Engineers
SAF	sustainable aviation fuel
SBTi	Science Based Targets initiative
SDF	Sustainable Development Fee
SDG	Sustainable Development Goals
SF-MST	Statistical Framework for Measuring the Sustainability of Tourism
SHA	Sustainable Hospitality Association
SIDS	Small Island Developing States
SMEs	small and medium-sized enterprises
SSMS	sustainable and smart mobility strategy
SSP	shared socio-economic pathway
SUP	single use plastic
tCO2e	tonnes of carbon dioxide equivalent
TCTF	The Conscious Travel Foundation
TEN-T	TransEuropean Transport Network
TIM	Travel Impact Model
TNC	The Nature Conservancy
TNFD	Taskforce on Nature-related Financial Disclosures
TPCC	Tourism Panel for Climate Change
TTC	Transformational Travel Council
UAM	urban air mobility
UATM	Urban Air Traffic Management
UGC	user-generated content
UN	United Nations
UNESCO	United Nations Educational, Scientific and Cultural Organization
UNFCCC	United Nations Framework Convention for Climate Change
UNGC	United Nations Global Compact
V2X	vehicle to everything
VFR	visiting friends and relatives
VIP	very important person

VR	virtual reality
WEF	World Economic Forum
WTTC	World Travel and Tourism Council
WWF	World Wildlife Fund
XR	extended reality
y-o-y	year-on-year
ZEV	zero emissions vehicle

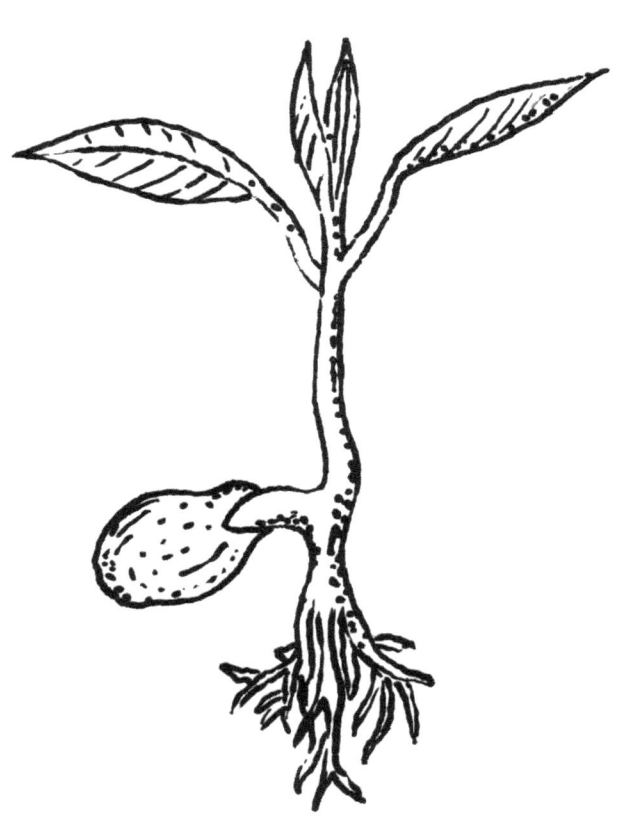

Introduction

Travel but at what cost?

Travel is ultimately a human desire. The ability to discover new places, cultures and enjoy new experiences. It can bring joy, awareness and connection. It forms part of people's complex needs to feel love, belonging, esteem and self-actualization. A multi-trillion-dollar industry has been built up over the past 200 years, turning billions of hopes, dreams and aspirations into reality enabling people to fulfil their passions. At best it is a force for positive transformation and empowerment of visitors and communities. At worst, it is extractive, destructive and perpetuates inequality and unfairness. Travel is therefore not a right – it is a privilege that comes with responsibility.

Strong consumer demand in line with growing disposable income, investment in transport infrastructure, connectivity, urbanization and digitalization have been key drivers in its growth. Digital transformation democratized travel at various junctures. The Global Distribution Systems (GDS) of the 1950s automated bookings, followed by low-cost carriers (LCCs) and online travel agents (OTAs) in the 1990s, opening up travel to the masses. Airbnb spearheaded the digitalization of short-term rentals in the late 2000s, flooding destinations with supply and exacerbating overtourism and causing housing crises as legislators played catch up.

Travel is regarded as an economic powerhouse and generator of hundreds of millions of jobs around the world, proclaiming itself a 'force for good'. Governments have fallen over themselves to offer tax incentives to encourage infrastructure, hotel developers, real estate investors and transport providers. Tourism helps countries to diversify economies away from traditional industries like agriculture or fossil fuels. Every corner of the world has seen tourism development. The rate of expansion has been at times excessive, leading to flashpoints worldwide where visitor capacity has been breached, entailing negative effects. From Venice, Barcelona, Dubrovnik, the

Canary Islands, Iceland, Boracay and Kyoto to Machu Picchu, there are bucket list destinations suffering from overtourism. While on the other hand, thousands of secondary and tertiary destinations are experiencing under-tourism with a lack of sufficient demand.

Yet at the heart of the travel industry resides an uncomfortable truth and dichotomy. The industry bears a significant carbon footprint, accounting for 8.1 per cent of the world's carbon emissions where transport and air travel in particular are called out for their high carbon intensity.[1] This puts travel under immense pressure to transition to a quality value-based tourism model that puts communities, biodiversity and the environment first.

Facing the worst-case scenario

The global pandemic signified the travel industry's worst-case scenario as it faced its greatest existential threat – the complete global shutdown of non-essential travel, tourism and hospitality. Climate change may not have directly shut down travel, but the linkages are evident. With zoonotic diseases at the nexus of climate change, ecosystem change and health, further global emergencies are expected as the climate emergency escalates.

In future, climate change will increasingly have direct impacts on travel. Already these effects are being experienced first-hand where peak temperatures, flooding, wildfires and extreme weather events are taking their toll on destinations and communities, and changing visitor behaviour.

Global targets, frameworks and pathways are in place to navigate the necessary reduction in global warming to 1.5°C compared to pre-industrial levels. Key milestones are earmarked for 2030 and 2050 in line with the United Nations Sustainable Development Goals (UN SDGs). With 2025 marking the year of peak global emissions, the writing is on the wall for carbon intensive industries such as travel. Sectors from airlines, airports, cruise and lodging, to booking platforms have united to commit to decarbonize and achieve net-zero emissions by 2050. Collaboration is seen as critical to ensure transparency and trust so that consumers can make the right choices when it comes to travelling.

Legacy challenges to overcome

Yet the travel industry, despite its claims to 'build back better' at the height of the pandemic, continues to power ahead with business as usual reinstated, with record-breaking numbers of trips and tourism spending.[2] Overtourism and anti-tourism sentiment or tourismphobia are back with a vengeance. Consumers, travel businesses and destinations are enjoying the 'last hurrah', continuing to travel with a *fin de siècle* mentality. Everyone knows that travel needs to not just change, but fundamentally transform. Still the growth at any cost mindset continues. Balancing the desire for growth with climate responsibilities is proving to be the paradox at the heart of travel. Decoupling economic growth from decarbonization has started but progress is slow.

There are many signs that there is a change in mindset, moving from words to action. Ever more destinations and travel brands are joining the Global Sustainable Tourism Council (GSTC) or signing up to the Glasgow Declaration for Climate Action in Tourism. Climate-positive travel brands are going one step further by becoming B Corp and/or adopting Science Based Targets Initiative (SBTi) targets. Targets require new metrics. The UN Tourism's Statistical Framework for Measuring the Sustainability of Tourism (SF-MST) aims to integrate new environmental and social metrics, looking beyond mere visitor numbers and spending. This marks a step in the right direction in assessing the true costs-benefits of travel and tourism, however, the framework has yet to be scaled globally.

Responding to the call for a paradigm shift

The transition from a volume-driven to value-driven tourism model is taking place. Moves include diversification of source markets, promoting longer lengths of stay, managing seasonality by promoting off-peak times, promoting less carbon intensive transport, and the equitable spread of tourism revenues through visitor dispersal. A whole suite of sustainable actions is being deployed to drive value to host communities by reducing revenue flight (so-called leakage), while promoting destination stewardship and accounting for the carbon footprint of tourism. This involves taking a holistic view of scope 1, 2 and 3 emissions across the supply chain.

However, actions are not going far, or fast enough. For an industry that has its foundations based on biodiversity, environment, culture, heritage and local communities, there is often a lack of coordinated action and urgency. Short-term gains are prioritized over long-term sustainable solutions at the mercy of disruptive forces like geopolitics, socio-cultural trends, urbanization, innovation and technology.

The first ever global stocktake of progress in travel and tourism to meet the 2030/2050 emissions pathways reveals that the industry is nowhere near where it should be. A 'paradigm shift' is needed to ensure success, not failure, where failure would be tantamount to destroying the planet for generations to come.[3] A global wake-up call is required with an open conversation about why all our travel behaviours should be more considered and impact conscious.

Fundamental transformation and wholesale change are required to future-proof travel. This process is going to be challenging, especially due to the highly fragmented nature of the travel industry, predominantly made up of micro, small and medium enterprises (MSMEs). Change needs to happen top-down as well as bottom-up at grassroots level. Equally important is how the travel industry talks about itself and demystifies opaque terminology so that impacts are fully understood. The narrative needs to change, moving away from the negative connotations of 'tourists', away from 'them and us'.

New business models are emerging in travel and tourism, stepping up to transform for the greater good. There is an evolution from sustainable to purpose-driven, regenerative, transformative, social impact, conscious, inclusive, redistributed and people-centric – all community-led. All these new business models ultimately embrace purpose: people, planet and profit for success. The destination marketing/management organization (DMO) of the past that focused solely on marketing is no longer fit for purpose as DMOs gravitate to holistic management first and foremost, with marketing as a secondary concern if at all.[4] Next-gen DMOs like New Zealand seek inspiration from indigenous knowledge and culture, while adopting the latest technology to future-proof their places and people. Others, like 4VI, are embracing a social enterprise model to ensure that profit is taken out of the picture to have a laser-sharp focus on positive impacts on local people, biodiversity and the environment for future generations.

Travellers onboard yet need to be nudged in the right direction

Despite public and private sector commitments, along with investor pressure, there is always doubt about whether consumers want to be or can afford to be more sustainable. Every year, surveys released by travel businesses like Booking.com report how consumers acknowledge the climate emergency and want to travel more sustainably. Many habits are already well established such as reusing towels or using paper straws. There is still plenty of room to accelerate adoption of more climate-friendly behaviours. These small incremental behavioural changes all add up, and consumers can be incentivized to make the better choice through gamification, rewards, education and information provided across the customer journey.

Consumers, especially Millennials and Generation Z, increasingly align their purchasing behaviour with personal values and beliefs. Travel businesses and destinations that prioritize and promote environmental and social benefits are likely to resonate and cut through the noise. Regulation such as mandatory sustainability reporting in the European Union (EU) is another reason to get ahead of the curve.

The future of travel demand lies in the emerging markets, especially Asia Pacific led by China and India. Travellers from emerging markets understand the real impacts of climate change, from extreme weather events and natural disasters to pollution. The future traveller will also be younger and digitally native. This requires a more inclusive approach that embraces diversity and planet-friendly travel experiences and services. There is another important dimension to the travel industry that is often ignored in that it facilitates the movement of migration. By 2050, there could potentially be 1.2 billion climate migrants and their movements should be accounted for in terms of building resilience.

If prohibitive measures are taken to curb travel and tourism due to climate change, it will be the future travellers from emerging regions that will bear the brunt, which is unfair and discriminatory. This impact will be two-fold as destinations will see job losses and residents will be unable to travel abroad. Efforts must be made to ensure that the transformation to net zero emissions leaves no one behind. Regions like the EU are leading the way in ensuring a just transition for all, not just their citizens.

'Travel is not just a part of the problem regarding the climate emergency, equity, and inclusion – it is part of the solution. In some places, it is the only solution so it needs to work well for everyone.'

Caroline Bremner, Author

Digital transformation – opportunities and threats

The travel industry is highly digitalized, especially for the search and inspiration phase of the customer journey and booking stage with 70 per cent of travel sales booked online in 2025.[5] Providing a seamless and safe travel experience is paramount to travel businesses. With advancements in cloud technology, artificial intelligence (AI), big data, IoT and biometrics, progress is being made to make the travel experience smoother. Travel facilitation benefits from the march of automation.

Blockchain and web 3.0 are being leveraged as travel businesses embrace the next generation of the internet. Further disruption is on the cards for commerce, payments, marketing and loyalty. With near ubiquitous internet and mobile connectivity, digital inclusion is rising. For example, Blockchain is being leveraged by local artisans to build a fair and transparent supply chain that promotes equity.

Digitalization drives personalization, traditionally in terms of tailored personalized offers and recommendations to the benefit of visitors. Personalized and curated travel services lead to higher value as the gap between expectation and experience is narrowed. This helps to drive quality tourism through added value, building a stronger connection between visitor and host community.

Embracing digital ID and digital inclusion can help to drive the social sustainability agenda, empowering local travel businesses, especially women that are the most representative in the workforce.[6] With 33 per cent of the world still offline, there remains some way to go so that digital inclusion is achieved globally.[7]

On the one hand, digitalization and automation will continue to influence travel operations, driving efficiencies, especially to reach net zero. On the other, automation represents an imminent threat. Gen AI is just at the beginning of its journey, with the jury out on how disruptive and harmful it will turn out to be. The bottom line is that automation leads to job losses, and this requires pre-emptive steps to reskill and upskill staff, removing the fear factor from making use of such tools. Governments need to get ahead of Gen AI so that its uses can be beneficial rather than destructive causing further bias and misinformation.

Further disintermediation is expected, in the way that OTAs disrupted tour operators and Airbnb disrupted lodging. Peer-to-peer business models using decentralized technology will unleash a new era of people-centric travel services with shared values and benefits for visitors and communities.

Boost to sustainable innovation

As target years of 2030 and 2050 approach, green technology and innovation will jump forward in leaps and bounds. Everywhere in the travel industry, the race to net zero emissions has started and integrated into long-term goals across multiple travel sectors.

Aviation, the hardest sector to decarbonize, is pushing on with its multi-fold approach, such as the transition to sustainable aviation fuel (SAF) and investing in electric/hydrogen aircraft. Finding new ways of travelling guilt-free is creating new start-ups especially in electric aircraft and electric vertical take-off and landings (eVTOLs). The roll out of electric vehicles (EVs) and EV infrastructure is becoming more integral to urban mobility solutions, and requires further scaling up and investment.

Airports and lodging players are investing in renewables, aiming to become net-positive energy hubs to give back to their communities. New partnerships with technology players lead to new collaborative efforts such as creating digital twins of hotels, airports and smart cities to make travel clean and seamless. Moving into Gen AI and the next-generation web 3.0 is helping to drive operational efficiencies and enhance the visitor experience. However, moving wholesale into such technology brings new types of challenges with higher computing power and carbon intensity.

Every travel business needs to take a holistic view of its negative impacts to remove them and maximize positive impacts on an ongoing basis. From waste, water usage and plastics to food sourcing, gender equality, inclusion, fair wages and positive social impacts all require constant attention.

Next-generation technology reshapes expectations

Travellers' expectations are set to change as new technology emerges. The roll out of 6G with even faster internet speeds will be a gamechanger. Augmented reality (AR) devices like Apple Vision Pro, followed by smart glasses, smart contact lenses or even wireless implants will lead into a new age of hyperconnectivity. The real and digital worlds will merge seamlessly due to omnipresent 6G wireless technology, expected in the 2030s.[8] With new technology capabilities with the Internet of Everything, travel businesses and destinations will face an uphill struggle to meet traveller expectations in real time. Partnerships with technology companies and start-ups driving transformation like Gen AI, web 3.0 and automation will be key. A proactive approach to technology and innovation, especially regarding nature-based solutions and community capacity building will be fundamental to success.

Futuristic transport such as next-generation supersonic aircraft travelling faster than the speed of sound are already being trialled such as Boom. Supersonic travel could well be commercialized again in the next decade. The business and luxury travel markets are poised to take advantage of faster, quieter flight times. Suborbital flights will be a possible alternative when travelling ultra long haul like Europe to Australasia in the next decade.[9] Autonomous vehicles (AVs) will move through their hype cycle to mainstream once deemed safe to roll out. All future transport solutions will need to be seen through the sustainability lens if they are to be developed and taken to market.

Future pathways to explore

For the majority of businesses, the future risks for travel are clear. The 2030/2050 framework already lays out what needs to happen when regarding a sustainable transformation to ensure a healthy planet for generations to follow. The direction of travel is clearly signposted. Alternative pathways are open but are closing fast as new tipping points are breached and global temperature records continue to be broken.

Yet the challenges are taking place on multiple fronts – social, environmental, biodiversity, digital, the workforce and diversity, equity and inclusion (DEI). Each destination and its communities face unique challenges and require a customizable approach to forge their own path within the global framework for what success looks like.

Failure to act together will lead to travel being cancelled, such as exclusive gated destinations, permanent visitor bans to iconic sites, prohibitive taxes, discriminatory visitor selection, flying only for the wealthy – just for starters. Potentially we will see the phenomenon of being 'loved to death' as the shutters come down on the most iconic natural and cultural sites in the world.

Instead, a collaborative and transdisciplinary approach is required: local communities, visitors, start-ups, private and public sectors, non-governmental organizations (NGOs), academia and all players across the supply chain. Travel needs to be plugged in with food, drinks, arts, music, culture, digital, architecture, mobility, energy, education etc. Success is a collective effort, led by communities working to secure their futures and homes in partnership (see Figure 0.1).

FIGURE 0.1 Travel transformation for a resilient and sustainable future

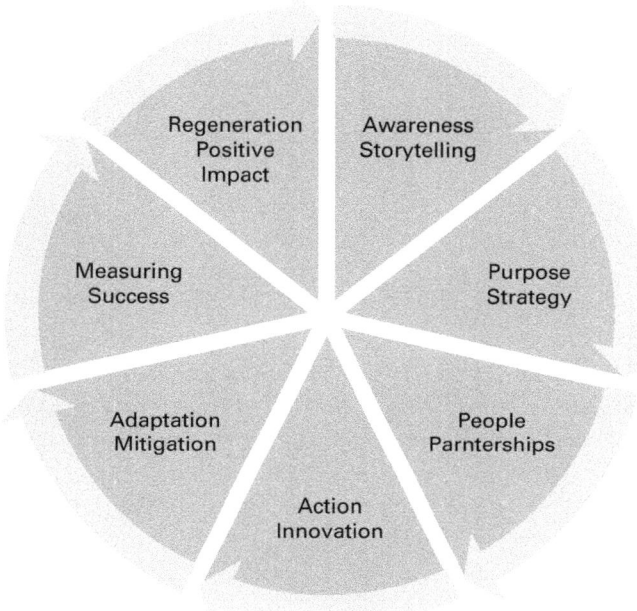

SOURCE Caroline Bremner, author

Shared future

At the heart of travel are the people, places, stories and ultimately our shared planet. The next generation of travel business models is already here, embracing net zero, regeneration, transformation and community tourism. By creating people-centric travel experiences and services based on the skills and needs of local communities, travel businesses can overcome the paradox that eats away at the industry. If not, some communities may turn their back on tourism if the exchange is not mutually beneficial.

Economic growth and jobs are important but need to be balanced with the cost-benefits of travel to communities, the environment and biodiversity. It is not a zero-sum game. The only way to future-proof travel for the next generations is to embrace transformative change whole-heartedly with eyes wide open. Armed with the latest insights, practical solutions and working in partnership will help forge the best path ahead with a sense of shared responsibility. Success is never guaranteed, but failure is not an option.

Endnotes

1 World Travel and Tourism Council (2022) wttc.org/news-article/wttc-unveils-industry-leading-and-ground-breaking-global-travel-and-tourism-sustainability-data-291122 (archived at https://perma.cc/WS77-GPVY)

2 World Travel and Tourism Council (2024) wttc.org/news-article/travel-and-tourism-set-to-break-all-records-in-2024-reveals-wttc#:~:text=As%20the%20global%20sector%20soars,underpinning%20almost%20348MN%20jobs%20globally (archived at https://perma.cc/LDW9-MMT5)

3 Tourism Panel for Climate Change (2023) Tourism and Climate Change Stocktake 2023 (Eds S Becken and D Scott), tpcc.info/stocktake-report/ (archived at https://perma.cc/6E23-Q7YU)

4 Solimar (2022) www.solimarinternational.com/why-dmos-must-be-about-destination-management/ (archived at https://perma.cc/KRE6-L38S)

5 Euromonitor International (2024) Travel 2025 edition

6 UN Tourism (2024) www.unwto.org/gender-and-tourism (archived at https://perma.cc/8JEP-5Y2Y)

7 World Economic Forum (2024) www.weforum.org/impact/digital-inclusion/ (archived at https://perma.cc/K7CR-B8XQ)

8 Ericsson (2024) www.ericsson.com/en/6g (archived at https://perma.cc/VSG3-EULL)

9 Australian Aviation (2023) australianaviation.com.au/2023/05/suborbital-flight-could-cut-london-sydney-trip-to-two-hours-within-10-years/#:~:text=Space-,Suborbital%20flight%20could%20cut%20London%2DSydney%20trip,2%20hours%20within%2010%20years&text=Suborbital%20travel%20could%20be%20accessible,Sydney%20in%20under%20two%20hours (archived at https://perma.cc/9N5B-PS99)

01

Travel at a crossroads

Introduction

The global travel industry stands at a crossroads, having faced its worst fear – the reality of zero non-essential travel as the pandemic brought tourism to a grinding halt across the world. The industry contemplated the biggest existential threat to its future survival to date. A threat even greater than the terrorist attacks of 9/11 or the Global Financial Crisis of 2008/2009 which had previously disrupted the industry's growth trajectory.

Governments closed their borders, imposed travel restrictions and confined people to their homes. For the majority, holidays were temporarily a dream, only played out online through virtual travel experiences. Yet there is an ever bigger existential threat that looms on the horizon: the triple planetary crisis of climate change, pollution and biodiversity loss.

To face future threats, industry leaders came together and resolved to 'build back better', promising to turn their backs on the old ways of volume-driven demand exemplified by overtourism and inequity. They promised quality over quantity. The pandemic was hailed as a once-in-a-lifetime turning point for a more sustainable and resilient tourism model that prioritizes communities, the environment and biodiversity over profit. The industry is making big strides in recognizing its carbon footprint and responsibility to make the transition to sustainable travel.

Post-pandemic, global visitor numbers and spending levels are breaking new records thanks to unprecedented pent-up leisure demand, marking a new era of growth. Consumers and travel businesses are enjoying the 'last hurrah', conscious that change is required to tackle the climate emergency, but not ready to adapt. Yet, in the new normal, the cracks in the current tourism model are clear to see. Overtourism and anti-tourism are back in popular destinations where demand is off balance and carrying capacity levels have been breached. Fundamental transformation is required to deliver the promise to guarantee long-term profitability that puts local communities and places first.

New threats and opportunities are emerging constantly. Next-generation technology like Generative AI and climate change are increasingly making their impacts felt; there is no room for complacency. The time for embracing sustainable and digital transformation is now. Businesses cannot work in isolation to tackle the challenges ahead due to the complexity of the supply chain with multiple stakeholders. Collaboration and partnership are critical for success. Already some are striking out with bold moves in climate action and social justice for tourism, sharing best practice for others to adopt and emulate.

Driver of economic growth and development

The success of travel and tourism is remarkable, from 700 million international arrivals in 2000, more than doubling in nearly two decades to reach 1.5 billion arrivals in 2024.[1] By 2030, the travel industry is forecast to exceed an impressive US$7 trillion in tourism spending, domestic and international combined, representing a major growth driver of the global economy.[2] The travel industry tends to outperform the global economy except in times of crisis. Over 2011–2019, real GDP growth rates in the global economy averaged 3.5 per cent compared to world international tourism spending of 5 per cent.[3, 4] With consumers increasingly looking for new experiences, there is a strong appetite for travel and businesses continue to see new opportunities.

The industry is identified by governments and investors as a robust driver of economic development and job creation. It is often the top revenue generator for countries, especially Small Island Developing States (SIDS) where often over 30 per cent of GDP comes from tourism.

The World Travel and Tourism Council (WTTC) showed that 1 in 5 new jobs were created in travel, while 1 in 10 jobs worldwide were directly or indirectly linked to tourism, amounting to 334 million in 2019.[5] By 2034, 449 million jobs worldwide are forecast to be derived from travel and tourism, despite accelerated automation, taking an even larger share of the global economy, at 11.4 per cent, amounting to an impressive US$16 trillion in direct, indirect and induced revenues (see Figure 1.1).[6]

FIGURE 1.1 World travel and tourism jobs and percentage economic contribution to GDP 2014–2034

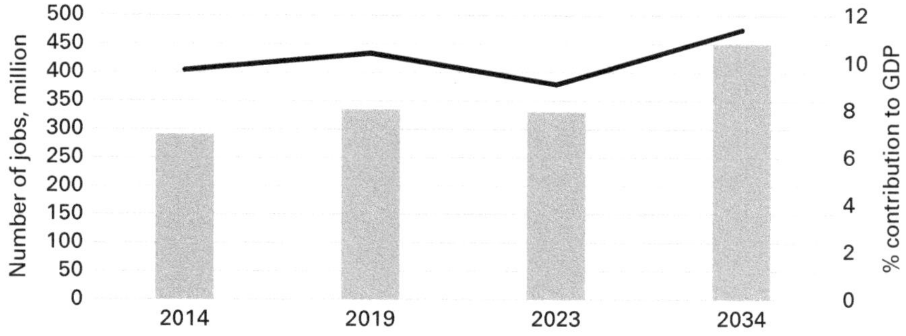

SOURCE WTTC, Oxford Economics[8]
NOTE Total contribution, including induced as well as direct and indirect contributions, 2034 forecast

Predominantly, travel and tourism are made up of micro, small and medium sized (MSME) enterprises. Of these, 80 per cent are mainly small family businesses, with a large percentage of youth and women employees, for example 54 per cent of tourism jobs are held by women.[7] The sector offers huge power in driving gender equality and equal opportunities such as access to education. The fragmented nature of the industry is a double-edged sword where travel and tourism support local businesses and communities and drive equality, however, it is difficult to drive change in a fast, coordinated and sustainable way. Seasonality challenges also effect businesses, leading to precarious and temporary work.

There is an enormous paradox: the travel industry has an extensive carbon footprint, with estimates of 8 per cent to 11 per cent of global greenhouse gas (GHG) emissions, in a range of 3.9 billion to 5.4 billion tonnes pre-pandemic.[9] Nevertheless, the industry continues to enjoy its role as a poster child of sustainable development, as a vehicle of robust economic growth and job creation, with the soft power gained from destination brand marketing. Countries powered by fossil fuels such as Saudi Arabia are undergoing a massive transformation to develop a tourism and leisure economy for a sustainable future.

Tourism is even called out in the United Nations Sustainable Development Goals (UN SDGs) as a means of achieving sustainable development, specifically, goal 8: decent work and economic growth; goal 12: responsible consumption and production; and goal 14: life below water.

Unprecedented declines in tourism expose vulnerabilities

Built on moving people safely around the world, the global pandemic represented the industry's worst fear: closure. Bans and restrictions brought an abrupt stop to all non-essential in-person activities. Travel, tourism, hospitality and leisure bore the full brunt of the pandemic restrictions, being the first to close and the last to reopen. Airlines and transport operators ground to a halt and laid off staff while some hotels pivoted to offering co-working spaces for remote workers, hosted hospital staff or repurposed their kitchens to provide food delivery services. Governments stepped in to provide financial support ranging from bailouts for airlines, business support schemes and fiscal stimulus amounting to billions of dollars.

At the height of the pandemic, international tourism spending dropped by 68 per cent compared to domestic tourism spending by 47 per cent in 2020 (see Figure 1.2). This compares to a drop of –2.8 per cent in world GDP, illustrating just how heavily impacted travel was compared to the wider economy and its decoupling from the economy.[10] For travel businesses and destinations hooked on growth, this was a devasting blow and a sign of potential challenges to come.

FIGURE 1.2 World inbound and domestic tourism spending 2019–2029

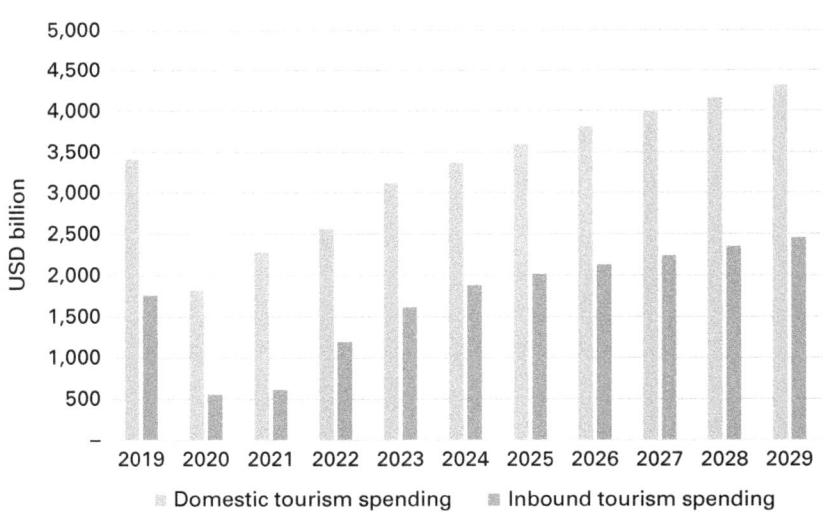

SOURCE Euromonitor International/World Tourism Organization (UN Tourism)
NOTE Value at constant 2024 prices, fixed exchange rates

Lessons learnt

The lessons of the pandemic are wide-ranging and will help shape future strategies for travel businesses in the decades to come. One critical learning point is the importance of collaboration and partnership with all stakeholders across the supply chain, especially in times of adversity.

Tourism boards, destination management organizations (DMOs), private sector – hotels, short-term rentals, transport providers, tour operators, food and beverage suppliers – formed coalitions to press for joined-up thinking and most importantly include local communities. This worked especially well at the national level, however, at a regional level it was very uncoordinated in regions such as Asia Pacific where introducing new travel protocols was disjointed.

Resilience from the staycation effect

Domestic leisure tourism globally proved to be a godsend for travel businesses and destinations once local travel restrictions began to ease enabling movement within a country. This quickly sparked a domestic staycation boom that companies such as Airbnb with its 'Go Near' campaign helped to ignite in Western Europe and North America. Consumers sought out off-the-beaten-track locations and wanted to get back to nature, rediscovering their freedom and passions. However, domestic tourism tends to be overlooked due to its lower spend per visitor than the high spending international visitor, despite being almost double the size in overall value (see Figure 1.2).

Digital acceleration

Following new health protocols introduced to share information on vaccination status for international travellers, a new era of digitization in travel was ushered in. This is exemplified by Covid-19 apps used in cases like the Dominican Republic or the EU's Digital Covid Certificate (DCC) deployed to facilitate travel during the pandemic.

Digital adoption accelerated as businesses pivoted online to engage with consumers. Some sectors like e-commerce thrived with online taking an ever greater share of retail sales. Travel businesses embraced online booking even more than ever, especially the less penetrated areas of experiences, tours and attractions to help with visitor flow management.

TIPS

- Importance of collaboration between the public and private sectors, across the entire supply chain to ensure working together.
- Improve scenario planning to ensure greater agility to respond to future crises especially climate events and natural disasters.
- More effective regional coordination of travel protocols.
- Greater emphasis on employee, customer and community wellbeing.
- Do not ignore the power of domestic tourism as a reliable source of revenue that can be leaned on in times of crisis.
- Be mindful that domestic visitors also have a carbon footprint.
- Designing and co-creating travel experiences with DMOs and visitors drives authenticity, loyalty and repeat visits.
- Travel businesses should continue to drive their digital transformation across the customer journey and achieve operational efficiencies.
- If businesses do not move with new technology like Generative AI, they will be left behind.
- Ongoing investment in digital skills is key for reskilling and upskilling staff.

Legacy labour shortages

The precariousness of travel and hospitality jobs is not new, often seen as low paid, with long hours and affected by seasonality. This is a reputation that is hard to break. Due to the severity of capacity slashed, the ability to scale up quickly after reopening was challenging for airlines, tour operators and hotels. Labour shortages continue to dog the industry, exacerbating labour disputes with management that lead to strikes. During the hiatus, many workers choose to seek alternative jobs rather than return due to perceived job insecurity. In the UK alone, the Office for National Statistics (ONS) reported that vacancies in hospitality stood at 6.6 per cent at the end of 2022, the highest of any sector.[11]

REAL-WORLD EXAMPLE
Marriott takes an inclusive approach to tackling labour shortages

Post-pandemic, Marriott reported that it was struggling to recruit 10,000 roles in the US as it faced a 'fight for talent' according to its Chief Executive Officer (CEO), Tony Capuano.[12] As hospitality reopened, furloughed staff were not tempted back, having moved into retail or e-commerce. Marriott, with 8,785 outlets, 31 brands across 138 countries, employing 411,000 people, is the leading global hotel company and prides itself on being a people-centric business.[13]

The company's recruitment and retention strategy evolves around the principles of growing great leaders, investing in associates and access to opportunities. Its diversity, equity and inclusion programme is company-wide, aiming to increase the representation of people of colour by 25 per cent at Vice President level and above by 2025.[14]

Rising costs of living and inflation combined with the labour supply shortages lead to greater pressure on hotel companies to raise wages. Consequently, Marriott highlights the risk to its business if it is unable to attract and retain associates due to competition from inside and outside the sector.

To help expand its talent pool, one major initiative is providing job opportunities for refugees and migrants, where it aims to hire 3,000 refugees worldwide by 2026.[15] It also is leading the hotel sector in combatting human trafficking by sharing best practice.

Marriott reported record results in 2023, with revenues of US$23.7 billion, up 14.2 per cent from US$20.8 billion in 2022.[16] Despite digitalization and accelerated automation, the role of people and creating an inclusive environment remain integral to Marriott's future success.

Challenges and threats in the new normal

Inflationary pressures

With peak inflation hitting record highs globally over 2022–2023, the resulting cost-of-living crisis adds to travel businesses' challenges, leading to labour tensions due to higher costs and pressure on consumer budgets. Across the supply chain, costs are unable to be fully absorbed and are passed onto businesses, ultimately to the consumer. Leisure travel holds the majority share of international tourism demand, however, it has a lower average spend per trip, making it more price sensitive. Yet travel businesses reaped the rewards from higher prices thanks to the uplift in demand from revenge travel (where people rushed to travel again after pandemic restrictions were lifted) and travel remains a top spending priority despite macro-economic headwinds. The danger is when consumers start to be priced out of the market especially as prices remain high and will rise as travel businesses increasingly pass on the costs of the sustainable transition.

Climate change in the spotlight

During the pandemic, the natural world flourished as society came to a standstill, with people confined to their homes and transportation paused. Social media about wildlife roaming free during the pandemic went viral including footage of dolphins in the canals in Venice.

For countries at the forefront of climate change, the realities of the climate emergency continue with no respite, and the world continues to experience the hottest temperatures ever on record every year. NASA reported that 2023 was the hottest year ever recorded, being 1.4°C warmer than pre-industrial levels.[17] In July 2024, the world experienced the highest global average temperature with more record temperatures expected to be broken and over June 2023–June 2024 temperatures were consistently 1.5°C above pre-industrial levels.[18]

For instance, the Arctic is warming three times faster than the rest of the world and experiencing devasting wildfires.[19] This leads to the impacts of climate change being felt worldwide as important natural sinks are unable to perform their role of carbon absorption. While SIDS are at the most risk from climate change, they are the least involved in creating it or responsible for it leading to climate injustice.[20]

Europe is experiencing climate change first-hand with wildfires, droughts and floods, and the increasing desertification of countries like Spain. The Mediterranean with its popular beach resorts is warming 20 per cent more than the global average.[21] Greece has suffered from extensive wildfires in destinations such as Rhodes that have caused death and destruction. The Greek National Tourism Organization has for several years been looking at ways to spread demand more evenly throughout the year, especially into shoulder seasons. Record temperatures will further influence traveller behaviour and travel businesses to look beyond peak summer months. So-called 'cool destinations' that enjoy cooler climates are gaining in appeal in peak summer months such as Scandinavia and the Baltics.

It is not just summer that is suffering from record temperatures in Europe as winter is equally seeing high temperatures. Many ski resorts have closed indefinitely or reduced their operating capacity. Ski tourism is expected to witness further closures, with an opportunity to develop adventure or wellness tourism, a diversification strategy already explored in Switzerland and Austria.

Elsewhere, increased temperatures are threatening to destroy coral reefs that are a vital ecosystem for biodiversity and popular with visitors for diving and wildlife watching. There are stark warnings that 90 per cent of coral reefs may die by 2050 if carbon emissions are not reduced urgently.[22] Often where there is coral bleaching, reefs are closed to divers as seen in Thailand, Malaysia and Vietnam over the past decade.

TIPS

- Spreading out seasonality and pushing demand throughout the year helps not just prevent overtourism and reduce climate risks, but attracts new traveller segments.

- Diversifying into passions and interests-based tourism is more price inelastic, opening up new sources of demand that are more resilient.

- Avoid promoting 'last-chance' tourism to see destinations or biodiversity at risk of extinction or that face irreversible changes if not conducted in a sustainable way.

- Travel to destinations at risk such as the Antarctic is responsible when driven by purpose, such as education, citizen science or conservation.

REAL-WORLD EXAMPLE
Fiji walks a fine line between economic, social and environmental needs

Fiji in the Pacific is highly dependent on international tourism revenues and suffered severe declines when tourism was shut down. Tourism accounts for a significant share of the country's economy at 40 per cent and its economy contracted by 17 per cent at the height of the pandemic.[23] It is a major source of employment accounting for 36.5 per cent of jobs in the country.[24]

The country is one of the most at-risk destinations from natural disasters and climate change impacts, which affects local communities, the food chain and its ability to earn tourism revenues. The country has joined forces with neighbouring

Pacific islands to vocalize the existential challenges they face due to their vulnerabilities to extreme climate impacts such as rising sea water and floods. They voice their concerns at global climate summits like COP about the threats facing their islands.

Over 99 per cent of visitors arrive by air according to Fiji Tourism.[25] Key source markets include Australia, New Zealand, China, Europe, the US and Canada. Dependency on air travel makes the transition to a sustainable tourism model challenging as it is one of the hardest sectors to abate for carbon emissions. Fiji Airways introduced sustainable aviation fuel (SAF) to blend with conventional fuel on its Airbus 350 from Singapore to Nadi (Fiji) in 2023. The airline reported that SAF produces 80 per cent less emissions than standard jet fuel, while being four to eight times more expensive.[26] Cost is not just the challenge for SAF but its limited availability, under 0.03 per cent of global fuel supplies.

With the support of the International Finance Cooperation (IFC)/World Bank, the country is following the Fiji National Sustainable Tourism Framework 2024 to 2033. The framework covers four goals: prosperous economies, thriving and inclusive communities, visible and valued cultures, and healthy islands and oceans. Aligned with the Global Sustainable Tourism Council (GSTC) criteria, the framework also follows the SAMOA Pathway for climate change and the Blue Pacific 2050 Strategy. In mid-2024, the World Bank provided financing of US$200 million to Fiji for the Nua Vualiku Tourism Development Project. The objective is to improve infrastructure and tourism in the northern island of Vanua Levu and serve as a model example of implementation of the Framework.

Post-pandemic, there is a shift to targeting higher quality visitors, seeking deeper immersion in Fiji's cultural and biodiversity offers. Adventure, sports, cruise, medical, meetings, incentives, conventions and exhibitions (MICE), wedding and eco-tourism are segments being developed alongside Indigenous and community tourism.

Balancing the needs of the community and the environment is tricky as shown by the FJD1 million fine imposed on a Chinese hotel and casino developer, Freesoul Real Estate, for causing damage to mangroves and the coral reef in Waci. The fine was applied under the Environmental Management Act where mangroves provide an important defence against climate change such as flooding and erosion.

RETURN OF OVERTOURISM AND ANTI-TOURISM

Overtourism is where visitor demand has tipped over into having negative impacts on a destination and the wellbeing of residents, causing social and environmental damage. The backlash to overcrowding and mass tourism practices is ever present across the entire travel ecosystem from travellers and DMOs to local communities fed up of seeing their places of residence flooded with too many visitors.

Local communities often complain about tourism being forced upon them with little say in how it is managed and the effects it has on their daily lives, environment, resources and biodiversity. Furthermore, while tourism supports the local economy, jobs in tourism are perceived to be low paid and precarious due to seasonality factors, leading to further frustration. The rise of short-term rentals thanks to the success of Airbnb has led to a housing shortage for residents that cannot afford to live near their place of work. In the Canary Islands, residents are joining together to protest against tourism, where they are asking for a moratorium on visitor numbers, an eco-tax and a change in the tourism model to quality not quantity.

The tide was turning even before the pandemic. Anti-flying movements such as flight-shaming from the Swedish word 'flygskam', promote travellers to choose alternatives to flying due to its higher carbon footprint. In tandem, the school strike for climate pioneered by Greta Thunberg quickly spread and drove awareness about climate impacts especially among young people.

In response, governments, destinations and travel brands are working to prove the value of tourism in terms of income generated locally, jobs created directly and indirectly, while taking a closer look at the use of natural resources where international visitors especially in emerging countries are much more carbon intensive.

TIPS

- Conducting regular surveys about resident satisfaction or creating citizen portals are a sure-fire way of keeping communities involved and feeling that they can have some influence over tourism development.
- Social listening techniques should be used to pick up early warning signs of overtourism.
- Collaborate with the next generation of eco-conscious social media influencers to promote responsible and sustainable travel behaviours.

EU GREEN DEAL LAYS THE GROUNDWORK FOR CHANGE

The EU, and the 27 countries it represents, is leading the green transformation with its EU Green Deal. Made up of different policies and the EU Climate Law, its ultimate goal is to transform societies and businesses to achieve carbon neutrality by 2050 and fight climate change. This is in order to achieve the goals of the Paris Agreement, by transitioning to the circular economy, smart mobility, greener energy and transportation, embracing the digital economy, adapting and mitigating climate impacts. There are many initiatives that fall under the deal including Fit for 55. Airlines and the maritime sectors are called out as needing to decarbonize and adopt innovative, green technologies. The EU Emissions Trading Scheme (ETS) is a mechanism to help with the transition, which includes sectors like aviation.

The EU is not just helping its own citizens, it aims to be a global leader in fighting climate change because climate change affects everybody. The Corporate Sustainability Reporting Directive (CSRD) is mandatory sustainability reporting in the EU for large and listed SMEs from 2025. Environmental, Social, Governance (ESG) criteria will be mandatory to disclose, including risks and the impacts that businesses have on the environment and societies. This includes mandatory reporting on metrics such as carbon emissions, waste, energy and gender equality. Micro businesses under 10 employees will be excluded; however, many are already pivoting to ESG reporting.

NECSTour – European Regions for Competitive and Sustainable Tourism – is working with the travel industry to help them make the necessary changes. Countries like Spain and Portugal that traditionally depended heavily on beach tourism have looked at ways to diversify their offer, tapping into rural tourism. Spain's National Rural Network platform aims to promote agritourism and rural tourism experiences.

Valencia is taking a holistic approach to visitor impacts, assessing the carbon footprint of tourism and detailing direct and indirect benefits to the destination, being the first to certify the carbon impact of tourism. Working with Global Omnium, the research presents the carbon impact of tourism, with 81 per cent of visitors' 1,268 million tons of CO_2 coming from transport including flights.[27]

Even before the pandemic, public opinion had turned on tourism. Anti-tourism movements and negative sentiment spread across Europe where southern European destinations especially have been at the forefront of overtourism. This prompted the European Parliament to review overtourism and propose policy responses to deliver greater balance for residents in overcrowded destinations.

However, overtourism is a global challenge. Boracay, a popular beach destination in the Philippines, was closed in 2018 by the national government for conversation purposes and reopened with a cap on visitor numbers. Iconic UNESCO World Heritage sites like Machu Picchu have imposed visitor limits and the need to apply for a permit to trek the Inca Trail for over two decades. This restricts numbers to 200 visitors and 300 porters and guides, not only creating a more unique experience for visitors but preserving the site.[28] In Japan, Kyoto banned visitors from parts of its centre due to overtourism concerns with overcrowding and visitor bad behaviour impacting negatively on the local community. Limits were also imposed on visitors to sacred Mount Fuji, almost halving numbers to 4,000 per day. These types of local counter-tourism interventions will become increasingly more commonplace to preserve cultural and natural assets.

REAL-WORLD EXAMPLE
Venice at the cutting edge of smart city solutions to combat overtourism

Venice is one of the most iconic cities in the world, deemed by UNESCO as an area of Outstanding Universal Value (OUV). It has attracted visitors for centuries with its rich cultural heritage, sites, museums, canals and bridges. Venice and its lagoon were nominated to the World Heritage List in 1987.

The city continues to struggle with the challenge of maintaining its way of life for residents, preserving the natural environment and infrastructure and avoiding depopulation as residents cannot afford to live in the centre. Cruise ships were found to be a major contributor to overcrowding and causing pollution, putting the World Heritage Site at risk of being moved to the endangered list. Bans on large cruise ships over 25,000 tons have been implemented in the San Marco basin – Giudecca Canal.

Globally, the city has been at the forefront of sustainable and digital transformation to ensure that it can be enjoyed for generations to come. The city was one of the first to introduce a smart control room to monitor, manage and control visitor and traffic flows in partnership with the council, TIM Group and police. The aim is to manage visitation, ensure minimal damage to the city and boost the quality of living for residents. The smart control room was opened in 2020, at the height of the pandemic, leveraging big data collected from Internet of Things (IoT) sensors, cameras and control units across the city, applying machine learning and artificial intelligence (AI) for predictive analytics, powered by 5G. The results are assessed in real time 24/7 by trained staff, illustrating that digital transformation does not preclude human involvement.

Venice also plans to introduce a new Access Fee of EUR5 in 2025 after a trial in 2024 for day trippers with talk of it doubling. Registration through QR code is required. The tax will be used to preserve the city and better manage visitor flows, relieving residents of paying for maintenance costs and waste management. Limits were introduced on large tour groups to 25 people with a ban on loudspeakers. Calls are being made to reduce short-term rentals to enable residents to live locally. The fight to regain control of Venice's sustainable tourism development continues.

TIPS

- Taxation alone will not solve overtourism and requires a holistic multi-pronged approach.
- In popular iconic destinations such as Venice, it is important to reconsider whether it is responsible to sell the destination especially during peak season.
- Looking for alternative second- and third-tier destinations, in emerging destinations, helps to relieve pressures.
- Refreshing the destination mix will open up new opportunities for travel businesses, communities and create new travel experiences.

Hidden costs

The Travel Foundation's 'Destinations at Risk: The Invisible Burden of Tourism' with Epler Wood and Cornell outlines all the hidden costs of tourism when it is not balanced in favour of local communities.[29] This seminal report highlights how traditional economic accounting fails to take account of the invisible costs of tourism that undermine the benefits at the destination and community level. These costs include the use of resources, infrastructure investment, debt, and cultural and natural assets' preservation. Social impacts are equally important to account for, such as the higher cost of living, shortage of housing and amenities for displaced residents and weakened community values. Failure by local authorities and travel businesses to get to the root of overtourism will further exacerbate the problems faced by communities as travel and tourism are set to continue on a growth trajectory.

Community direct action

Some communities end up taking direct action to preserve the integrity of their places of residence and work, coming together to provide a united front against overtourism. Particularly in southern Europe, the negative resident sentiment is visible through anti-tourism graffiti, protests and tourism-phobia attitudes such as referring to invasions or hordes. Cities such as Venice, Barcelona and the Canary Islands have been flashpoints for anti-tourism protests. In County Clare, Ireland, local residents previously blocked access to routes taken by tour buses. These manifestations of antitourism may lead to new policies and taxation so travel businesses should be mindful of such localized actions of resistance.

However, as Scott Wayne of SW Associates, LLC – Sustainable Destinations explains, for every anchor destination that is overrun with visitors where capacity limits have been breached, there are hundreds of other smaller destinations experiencing under-tourism, especially in emerging and developing destinations. Secondary and tertiary destinations are crying out for visitors. Under-tourism is as much part of badly run destination management as overtourism where dispersal has not been set up correctly.

Taxation as a lever of destination management

Countries around the world are increasingly looking at 'tourist taxes' not just to raise revenue but to invest in the preservation and future development of places. The Visitor Levy bill in Scotland passed in 2023 will enable local councils to introduce levies if deemed necessary where the amount will be a percentage of the accommodation cost. Edinburgh will be the first city in Scotland to introduce a tourist tax at 5 per cent of the daily bed night surcharge.

TIPS

- Flexing taxation to deter and/or encourage visitors is not a panacea for attracting the right type of visitor, but it can help manage tourism flows more efficiently and fund investment in much-needed housing and infrastructure.

- One size does not fit all, where taxation in one location may be required where overtourism exists, but it may be prohibitive to under-supplied destinations.

- Flexibility is key along with frequent reviews and adopting an agile approach to deployment.

- Transparency of where money raised through taxation is spent locally is important so that residents and visitors understand the benefits of such schemes.

BHUTAN CHANGES TACK ON TAXATION

Pre-pandemic, Bhutan took a strong stance in ensuring that the country's mountainous landscape and environment were protected from the negative effects of mass tourism. It charged a Sustainable Development Fee (SDF) to international visitors, originally at US$200 per day. This clear labelling and designation of funds for conservation and preservation was successful in deterring budget travellers, leading to a selective focus on luxury travellers.

Post-pandemic, the government decided to lower the rate by 50 per cent to US$100 per day in 2023 for a period of four years, to encourage demand, with the rate capped for longer stays.[30] Striking the right balance between visitor demand and spending is challenging. Bhutan aims to target MICE tourism where blended trips including a leisure add-on will be encouraged by the changes in fee.

TIPS

- Taxation is a useful lever for raising revenues to preserve local sites and heritage, however, it can be off-putting to consumers unless the link is clear between taxation and sustainable development.
- A flexible approach is required to taxation that it can be flexed across different destinations and seasons.
- As with any sustainability claims, these must be backed up with concrete evidence and hard data to avoid greenwashing.

New normal

From words to action

Travel leaders from the public and private sectors are rising to the challenge, working with the likes of NGOs specializing in sustainable tourism like the Global Sustainable Tourism Council (GSTC), The Travel Foundation, the Future of Tourism Coalition, and global organizations like UN Tourism and WTTC. There are over 470 active travel and leisure participants in the UN Global Compact, which has a voluntary scheme to follow sustainability principles and responsible business practices in line with the SDGs.[31]

At the height of the pandemic, the Future of Tourism Coalition was formed, promoting 13 guiding principles for travel. The core pillars focus on climate action, measurement and metrics, supply chain and destination stewardship. With the pandemic allowing destinations time to reflect on the loss of tourism, travel brands and destinations agreed that it was time to change to ensure long-term survival. There was a general consensus to transform to a quality-driven tourism model with SDG goals in mind.

FUTURE OF TOURISM COALITION'S 13 GUIDING PRINCIPLES

1 See the whole picture

2 Use sustainability standards

3 Collaborate in destination management

4 Choose quality over quantity

5 Demand fair income distribution

6 Reduce tourism's burden

7 Redefine economic success

8 Mitigate climate impacts

9 Close the loop on resources

10 Contain tourism's land use

11 Diversify source markets

12 Protect sense of place

13 Operate business responsibly

SOURCE Future of Tourism Coalition (futureoftourism.org)

TIPS

- With so many different membership organizations and accreditation schemes, it is difficult to see the wood for the trees.

- With global standards from the UN's Statistical Framework for Measuring the Sustainability of Tourism (SF-MST) programme not yet in place, many destinations remain in limbo.

- Additional resources on useful organizations to get helpful information from and partner with are available in the Additional Resources section at the end of the book.

Tourism officially declares a climate emergency

At COP26, the Glasgow Declaration on Climate Action in Tourism was announced in 2021, with travel businesses and destinations encouraged to declare a climate emergency. This involves signing up to deliver a climate action plan and enact this within one year of signature to halve carbon emissions in line with the Paris Agreement by 2030 and reach net zero before 2050. The declaration encompasses five pathways to sustainable tourism: measure, decarbonize, regenerate, collaborate and finance. Among the first to sign included Scotland as the host of COP26 where signatories span the gamut of travel businesses and destinations.

Travel businesses continue to sign the Glasgow Declaration, partner with NGOs, and adopt ESG reporting and sustainability practices such as by going B Corp. One step further is committing to the Science-based Targets Initiative (SBTi) that puts science at the heart of business practices. The SBTi works with companies and organizations across industries, where they publish their targets and level of commitment to achieving net zero by 2050. Intrepid Travel is a global leader in adventure travel that is B Corp certified and part of the SBTi where the two are highly complementary. Intrepid Travel is also a long-standing participant of the UN Global Compact (see Table 1.1).

TIPS

- Adopting sustainability criteria and following the science are the new business norm.
- With mandatory ESG reporting for large companies, this will help pave the way for shared best practices for all.
- Finding like-minded partners and organizations will help accelerate the required transition over the short, mid and long term.
- Social impact is as vital as environmental impact for inclusion and diversity.

TABLE 1.1 Select travel businesses' climate action targets

Category	Company Name	Near term - Target Classification	Near term - Target Year	Net-Zero Committed	Target Details
Airlines	Air France - KLM Group	Well below 2°C	FY2030	No	Commits to reduce well-to-wake scope 1 and 3 jet fuel GHG emissions 30% per revenue tonne kilometer (RTK) by FY2030 from a FY2019 base year
Airlines	Air New Zealand	Well below 2°C	2030	No	Commits to reduce well-to-wake GHG emissions related to jet fuel 28.9% per revenue tonne kilometre (RTK), equivalent to a 16.3% absolute reduction, by 2030 from a 2019 base year
Airlines	American Airlines	Well below 2°C	2035	No	Commits to reduce well-to-wake GHG emissions related to jet fuel 45% per revenue ton kilometer from owned and subcontracted operations by 2035 from a 2019 base year
Airlines	ANA Holdings Inc.	Well below 2°C	FY2030	No	Commits to reduce scope 1 and scope 3 fuel and energy related activities GHG emissions 29% per RTK by FY2030 from a FY2019 base year
Airlines	Delta Air Lines	Well below 2°C	2035	Yes	Commits to reduce well-to-wake scope 1 and 3 jet fuel GHG emissions 45% per revenue tonne kilometer by 2035 from a 2019 base year
Airlines	easyJet plc	Well below 2°C	FY2035	Yes	Commits to reduce well-to-wake GHG emissions related to jet fuel 35% per revenue tonne kilometre by FY2035 from a FY2019 base year

(continued)

TABLE 1.1 (Continued)

Category	Company Name	Near term - Target Classification	Near term - Target Year	Net-Zero Committed	Target Details
Airlines	Lufthansa Group	Well below 2°C	2030	Yes	Commits to reduce its well-to-wake GHG emissions related to jet fuel from owned operations 30.6% per revenue tonne kilometre (RTK) by 2030 from a 2019 base year
Airports	Royal Schiphol Group	1.5°C	2029	Yes	Commits to reduce absolute scope 1 and 2 GHG emissions 46% by 2029 from a 2019 base year. Also commits to reduce absolute scope 1 and 2 GHG emissions 90% by 2050 from a 2019 base year
Intermediaries	Intrepid Travel	1.5°C	2035	Yes	Commits to reduce absolute scope 1 and 2 greenhouse gas emissions 71% by 2035 from a 2018 base year. Also commits to reduce scope 3 greenhouse gas emissions from its offices by 34% per full-time equivalent and from its trips by 56% per passenger day over the same period
Intermediaries	The Travel Corporation	1.5°C	2030	Yes	Commits to reduce absolute scope 1 and 2 GHG emissions 46.2% by 2030 from a 2019 base year. The Travel Corporation also commits to reduce absolute scope 3 GHG emissions 27.5% by 2030. Commits to reduce scope 1, 2 and 3 GHG emissions 90% by 2050 from a 2019 base year

Sector	Company	Temperature	Target year	Validated	Commitment
Intermediaries	TUI Group	Well below 2°C	FY2030	No	Commits to reduce absolute scope 1 and 2 GHG emissions from hotels 46.2% by FY2030 from a FY2019 base year. TUI commits to reduce scope 1 and scope 3 well-to-wake emissions from jet fuel 24% per revenue passenger kilometer by FY2030 from a FY2019 base year. TUI commits to reduce absolute scope 1 and scope 3 well-to-wake emissions from marine cruise fuel 27.5% by FY2030 from a FY2019 base year. The target boundary includes land-related emissions and removals from bioenergy feedstocks.
Lodging	Accor S.A.	1.5°C	2030	Yes	Commits to reduce absolute scope 1 and 2 GHG emissions 46% by 2030 from a 2019 base year. Accor also commits to reduce absolute scope 3 GHG emissions from purchased goods and services, fuel and energy-related activities and franchises 28% over the same timeframe
Lodging	Airbnb, Inc	1.5°C	2030	No	Commits to reduce absolute scope 1 and 2 GHG emissions 78.4% by 2030 from a 2019 base year. Airbnb, Inc. further commits to reduce scope 3 GHG emissions 55% per M USD of gross profit by 2030 from a 2019 base year
Lodging	Hilton	1.5°C	2030	No	Commits to reduce absolute scope 1 and 2 GHG emissions 46.2% by 2030 from a 2019 base year. Hilton also commits to reduce absolute scope 3 GHG emissions from franchises 27.5% within the same timeframe

(continued)

TABLE 1.1 (Continued)

Category	Company Name	Near term - Target Classification	Near term - Target Year	Net-Zero Committed	Target Details
Lodging	Hyatt	Well below 2°C	2030	No	Commits to reduce absolute scope 1 and 2 GHG emissions 27.5% by 2030 from a 2019 base year. Commits to reduce scope 3 GHG emissions by 53% per square meter by 2030 from a 2019 base year
Lodging	Iberostar Hotels and Resorts	1.5°C	2030	Yes	Commits to reduce absolute scope 1 and 2 GHG emissions 85% by 2030 from a 2019 base year. Also commits to reduce absolute scope 3 GHG emissions
Lodging	InterContinental Hotels Group PLC	1.5°C	2030	No	Commits to reduce absolute scope 1, 2 and scope 3 GHG emissions from fuel and energy-related activities and franchises 46.2% by 2030 from a 2019 base year
Lodging	Las Vegas Sands Corp	Well below 2°C	2025	No	Commits to reduce absolute scope 1 and 2 GHG emissions 17.5% by 2025 from a 2018 base year
Lodging	Melia Hotels International SA	1.5°C	2025, 2035	No	Commits to reduce absolute scope 1 and 2 GHG emissions 29.4% by 2025 and 71.4% by 2035 from a 2018 base year. Also commits to reduce absolute scope 3 GHG emissions 29.4% by 2025 and 71.4% by 2035 from a 2018 base year
Lodging	Millennium & Copthorne Hotels plc.	Well below 2°C	2030	No	Commits to reduce absolute scope 1, 2 and 3 GHG emissions 27% by 2030 from a 2017 base year
Lodging	NH Hotel Group	2°C	2030	Yes	Commits to reduce absolute scope 1, 2 and 3 GHG emissions 20% by 2030 from a 2018 base year

SOURCE Science-based Targets Initiative[32]

Measure for success

Transformation can only begin with a solid understanding of a travel brand's or destination's starting point, which begins with a baseline measurement. There are hundreds of certification schemes for travel and tourism including EarthCheck, Travelife, Green Key, EU Ecolabel and Green Tourism among the top. These schemes enable travel businesses or destinations to receive third-party certification about their sustainability credentials which consumers value. For example, 65 per cent of Booking.com's consumers would feel better about staying in a certified property if it had a sustainable certification or label.[33] However, with so many schemes it can be confusing for consumers and businesses.

GSTC is responsible for the global standards for sustainable tourism, provides accreditation for certifying bodies while it does not conduct certification directly and recognizes where standards are equivalent to GSTC criteria. The GSTC criteria relate to destinations and the private sector like hotels, tour operators and MICE, with attractions in the pipeline. The GSTC criteria are all linked to the 17 SDGs that make up the Sustainable Development 2030 agenda, setting the minimum for organizations to aim for. The criteria follow the industry-wide ISEAL Alliance standards, that are linked to ISO standards where every company and location are unique, and the criteria should be adapted to their specific circumstances.

The four main pillars centre around sustainable management, socio-economic impacts, cultural impacts and environmental impacts, adapted to the type of organization. With changes in sustainability reporting in regions like Europe, GSTC is well positioned to lead the industry on standards and reporting, providing third-party impartiality to ensure that consumers can trust in green labels. For example, countries like Singapore have become GSTC destinations certified through Vireo Srl.

GSTC DESTINATION CRITERIA

Sustainable management

- management structure and framework
- stakeholder engagement
- managing pressure and change

Socio-economic sustainability

- delivering local economic benefits
- social wellbeing and impacts

Cultural sustainability

- protecting cultural heritage
- visiting cultural sites

Environmental sustainability

- conservation of natural heritage
- resource management
- management of waste and emissions

SOURCE Global Sustainable Tourism Council

TACKLING OVERTOURISM

Identifying the root causes of overtourism is fundamental to finding the right destination management solutions. These factors tend to have built up over many years, where tourism was pinpointed as an economic driver, with millions of dollars poured into marketing the destination. This goes hand in hand with airport expansion, air connectivity and wooing new airlines and routes to deliver visitors. Resort development and real estate investments are then the next stage in the process, often with governments offering attractive tax incentives.

Spurred on by digitalization, global online travel agents (OTAs), low-cost carriers, Airbnb and short-term rentals have helped to democratize travel, making it more affordable. Combined, increased connectivity, marketing and social media have led to a golden age of travel, particularly in the 2000s. Yet each destination has its own limits. Where overtourism thrives is where carrying capacity has been breached. Airbnb as the leading short-term rental player is often in the headlines and at loggerheads with local legislators to ensure housing remains available for residents and tourism workers. However, this is often not the case and holiday rentals are often cited alongside overcrowding as the reasons for anti-tourism sentiment.

Destinations suffering from overtourism such as Venice, Barcelona and Dubrovnik have also pinpointed cruise ships as a key contributor to excessive crowds where visitors jostle with residents. When asked which type of visitor brings the least value to their city, Dubrovnik residents highlighted cruises as the least to contribute. Amsterdam became another city turning its back on cruise visitors in 2023 to limit visitor numbers and curb pollution.

Dubrovnik featured in the global hit series, *Game of Thrones*, along with Iceland and Northern Ireland. The Netflix effect can catapult destinations into the global media spotlight, adding filmset locations to fans' bucket lists, topping Instagram and TikTok and amplifying the allure. It is hard to put the genie back in the bottle when it comes to 'set jetting'.

The 'bucket list' is a concept that continues to inspire consumers. One way to help take the pressure off iconic destinations like Bali is by offering alternative places to explore, encouraging visitor dispersal. The Indonesian government is taking the heat off Bali, by creating 'five super priority destinations'. Likewise, Peru has been promoting alternative trails to the Inca Trail.

REAL-WORLD EXAMPLE
Dubrovnik pulls out all the stops to preserve its city and quality of life for residents

The historic city of Dubrovnik in Croatia on the Adriatic, famed for its fortified city walls, is protected by UNESCO. However, for years it has struggled with overtourism, detrimental not only to residents but also its natural and cultural assets. This led to the city's Integrated Action Plan in partnership with the EU's URBACT network that includes nine cities including Venice and Krakow with the aim to promote sustainable tourism.

What is impressive about Dubrovnik's approach is the shift away from sector-specific actions to multi-stakeholders across the private and public sectors where negative tourism impacts require a holistic and collaborative approach. The top negative impacts identified were congestion, pollution and lack of parking spaces.

While the top three challenges were defined as the lack of cooperation across sectors, lack of understanding about what sustainability is and the uniform and profit-driven tourism offer being promoted. Key goals include expanding the visitor season, developing more original visitor experiences while improving transport and cultural conservation.

Cruise plays a major role in the city's tourism revenues but equally has caused overcrowding, bottlenecks and congestion. Cruise passengers arriving for the day do not bring value to the city due to the low average spend. To move away from uncontrolled development, the city is working with the GSTC where traffic management, waste management and support for local businesses were pinpointed as key focus areas. Caps were introduced in 2019 to half the number of cruise passengers to 4,000 entering the city at one time, restricting the number of berths per day to two at staggered times. Yet challenges remain as record visitor numbers continue, exceeding 4 million in 2023.[34]

Adopting a multi-modal transport strategy is central to managing the flows of visitors and residents especially in the area of Gruz where the port and main transport hubs are located. Other key actions include restricting tour buses from entering the old town and encouraging the use of public transport with the addition of electric buses and online ticketing. Other measures to reduce congestion included removing souvenir stalls and reducing tables and chairs at restaurants.

Smart city initiatives such as using an app, machine learning and AI for its visitor demand management system have been rolled out. Environmental monitoring sensors have been introduced to measure a wide range of factors from light and noise to air pollution. Cultural heritage has been digitalized such as providing VR tours in museums and information around town accessible via QR codes.

City of Dubrovnik Development Agency (DURA) is responsible for the measurement and reporting of the city's progress in transforming to sustainable tourism that delivers quality of life for residents and enjoyable visitor experiences.

Zero tolerance

Some destinations are taking a zero-tolerance approach to unsustainable tourism and bad visitor behaviour to ensure that the quality of visitation aligns with the way of life of residents. This involves pre-emptive destination marketing to put off certain groups from travelling. In tandem, a highly sophisticated destination management strategy is deployed that uses cutting-edge technology, takes account of residents' concerns and preserves the integrity of the place and culture. Cities like Amsterdam are leading the transformation away from destination marketing to destination management, and the next generation of DMOs focused on management.

In Bali, the local government announced a ban on sunrise hikes to sacred mountains due to visitors' bad behaviour but this was quickly lifted due to local concern about job losses. This shows the complexity of striking the right balance for everyone involved in the travel ecosystem.

TIPS

- A zero-tolerance approach can work if all stakeholders are part of the destination consultation process.
- One way to ensure all visitors know what is expected of them is to ask them to sign a pledge to respect people and place as seen in Iceland and Palau.

- Identifying key traveller segments with shared values will help to align visitor and community expectations.
- Working with companies like Airbnb that rate guests and not just hosts is one way of identifying potential bad behaviour.

REAL-WORLD EXAMPLE

Amsterdam takes bold action and bans destructive tourism

Amsterdam is one of the top European cities to visit with its unique attributes from its liberal freedoms, canals, world-class museums, art galleries and buzzing nightlife, welcoming 9 million international visitors in 2023.[35]

The city has struggled with balancing the needs of residents and businesses with the large numbers of visitors in the city centre, traditionally popular with hen and stag parties. After the pandemic, the city government took the bold step to limit visitor numbers and stop tourism activities that focus on alcohol, sex and drugs. These restrictions and looking for alternative sources of demand is pivotal to the success of the city's tourism strategy to 2035 to ensure quality of living standards are met by transforming into a sustainable visitor economy.

Key goals are to transform the city centre back to being a place to live and work for residents, rather than overnight lodging. Limiting and stopping certain types of visitors, along with dispersal beyond the city, means shifting away from destination marketing and piling all efforts into destination management. The city has been highly proactive in keeping a check on the proliferation of short-term rentals by introducing new regulations like the need for a permit.

The city is a global leader in saying enough is enough, adopting a peak tourism strategy. In the same vein as IKEA calling time on 'peak stuff', Amsterdam has called time on destructive tourism with no limits. It is not just a case of adjusting demand. The city is taking a strict approach to supply, promoting responsible businesses with viable and sustainable revenues that contribute in a positive way to the visitor experience and way of living.

Its vision for 2035 is respectful visitors from all over, of all ages. Dispersal to other regions will help minimize pressures in Amsterdam while investing in new cultural centres. Budget flights will be swapped for rail. Bans on new hotels and short-term rentals controls will continue to be enforced. The city's new tourist taxes introduced in 2024 are the highest in Europe, applied to cruise and hotel visitors. Example measures introduced to prevent bad tourism behaviour include restrictions on stag parties, hen parties and pub crawls, earlier closing times and bans on smoking cannabis on the street.

TIPS

- There is an acceptance that degrowth is not a viable strategy as destinations want to remain open to cultural exchange and visitors from all corners of the world.

- Taking a proactive approach that puts communities, residents and place first will help move the needle.

- Closing off unsustainable sources of demand is a wise move that will lead to better outcomes in the long run.

Evolving business models

Regenerative tourism

One of the buzz words in travel is regenerative tourism. Regenerative tourism ensures that tourism gives back more to people and places than it takes. Its aim is to provide a net positive effect on people, places and nature for long-term renewal. Destinations like New Zealand and Flanders, working with regenerative tourism leaders such as Anna Pollock of Conscious Travel, became the guiding lights for regenerative transformation at the height of the pandemic. The model takes a holistic view of the planet, taking inspiration from indigenous knowledge and the interconnectedness of life on earth.

REAL-WORLD EXAMPLE
New Zealand spearheads the transition to regenerative tourism

The government of New Zealand created a Tourism Futures Taskforce to work on the masterplan for regenerative tourism led by four wellbeing principles: economy, nature, culture and society, targeting high-value visitors. The country's tourism policy is informed and inspired by its indigenous Māori culture and wisdom. The aim is to drive the positive benefits of tourism as the country's largest export industry. In 2023, travel and tourism contributed 6.2 per cent to the New Zealand economy in direct and indirect contributions.[36]

Celebrating and preserving Māori culture is central to its strategies, marketing and unique visitor experiences, to create emotional connections with *manuhiri* (the Māori term for visitors).

The regenerative tourism approach demands active engagement from visitors, who are encouraged to take the Tiaki Promise (meaning to care for people and place) and see the world through this indigenous lens. The tiaki symbol and its design reflect the four spiritual beings that represent four main elements: sky, forest, earth and ocean.

Such strong emphasis on delivering positive benefits for future generations based on the culture and heritage of the past, elevates the connection between visitors and communities. This engenders a shared responsibility in caring for people and place that will stay with visitors after they leave.

New Zealand Aotearoa: Tiaki Promise[37]

- Care for land, sea and nature, treading lightly and leaving no trace.
- Travel safely, showing care and consideration for all.
- Respect culture, travelling with an open heart and mind.

FIGURE 1.3 Tiaki Promise

SOURCE New Zealand Tourism

Social impact

A new generation of DMOs are beginning to emerge, removing the influence of profit and putting the emphasis firmly on climate action, social justice, biodiversity and cultural preservation. Through a non-profit and purpose-driven lens, these organizations work with the private and public sectors to drive transformative and regenerative impacts at the local level. Collaboration is key as not one organization can achieve the necessary system-wide changes without partnership.

Fairbnb is a fine example of a community-powered travel cooperative, driven by a purpose to deliver the best outcomes for local communities as well as authentic experiences for visitors; 50 per cent of the company's booking fee for short-term rental bookings funds local community projects. The company takes an active stance to not promote any form of mass tourism, instead promoting responsible, fair and circular practices. As a cooperative it is people-powered and people-centric, for ensuring the most socially sustainable results to redistribute and share prosperity. Its new FairUp experiences platform uses local scouts to promote underrepresented businesses to ensure tourism revenues stay local.

REAL-WORLD EXAMPLE

4VI – New generation regenerative tourism business, focused on climate action and social justice

When it comes to the next generation of DMOs, 4VI is hailed as a leader. Representing the interests of Vancouver Island, the organization has transformed into a not-for-profit social enterprise. Its primary goal is to promote sustainable tourism and mitigate the negative impacts of tourism and climate change in keeping with the 17 SDGs. It works in partnership with multiple stakeholders including indigenous communities with a focus on capacity building and education where it considers regenerative tourism and indigenous tourism to be intertwined.

The organization is well accredited, as a signatory of the Glasgow Declaration, the Rainbow Foundation to promote LGBTQ+ rights and it received a Biosphere Certificate from the Responsible Tourism Institute in 2022. In keeping with the Glasgow Declaration, the goal is to halve emissions by 2030 and achieve net zero ahead of 2050.

Its climate action plan focuses on five pathways: measure, decarbonize, regenerate, collaborate and finance, where each pathway has its own set of goals and objectives. Fundamental to success is understanding the carbon emissions from tourism with a baseline of emissions intensity per visitor of 193 kg CO_2e, based on

10.2 million visitors, producing nearly 2 million tonnes of carbon.[38] By 2030, 4VI aims to reduce the average carbon footprint per visitor to 100 kg CO2e, working with 150 travel businesses and 10 Community DMOs.

Transport was the biggest contributor to emissions at 51 per cent, of which international flights accounted for 69 per cent, while food, drinks and retail were the second largest contributor at 36 per cent, compared to 11 per cent for lodging.[39] Long-haul source markets Australia and China were called out as having the highest carbon footprint vs domestic visitors.

4VI is at the forefront of climate action and social justice to drive the necessary changes required for a wholesale sustainable and resilient transformation in tourism. From micro-finance to nature-based solutions, it is pioneering a new generation of purpose-driven DMOs, where taking profit out of the equation helps to provide clarity of purpose to the entire ecosystem.

Joining forces with like-minded partners

There are many organizations that bring together like-minded people who care about people, the planet, biodiversity, conservation and ultimately having a positive impact.

The Long Run is such a movement, working with sustainable and nature-based tourism businesses with its 4C framework: Conservation, Community, Culture and Commerce. Each member aims to be awarded Global Ecosphere Retreats status, aligned with GSTC criteria.

The Conscious Travel Foundation (TCTF) is another social enterprise helping members to achieve positive impacts, act with purpose and drive sustainable, responsible and inclusive tourism with ambassadors promoting accessibility and DEI. In its Impact Report 2022, the TCTF encompasses 22 countries and 55 members. Members include IncluCare, the world's first guest-inclusion verification for luxury hospitality, to bring true inclusion to travellers with hidden or non-hidden disabilities.

Regenerative Travel brings together those travel businesses aiming to give back more than they take, featuring members from independent hotels and travel designers offering farm stays to mountain retreats. Travel by B Corp brings together UK travel businesses that have joined B Corp to adhere to legally binding environmental and social criteria. Key members include Intrepid Travel, Exodus Adventure Travel, Much Better Adventures, Pura Aventura and Journeys with Purpose, among others.

Evaneos, a French-based B Corp, is highly engaged with the 2030 climate action agenda. Key strategies include the removal of less-sustainable products such as city trips of less than five days and promoting rail travel and shorter-distance trips. It co-creates trips with its travellers based on shared values of alternative experiences, ecosystem regeneration in partnership with communities, cultural exchange and independently owned accommodation.

Transformational travel

Transformation is another buzz word in travel and has multiple meanings. For the Transformational Travel Council (TTC) it is about inspiring the next generation of travellers through mindful, conscious and regenerative tourism. The Council benefits from the experience of hospitality guru, Chip Conley, as an adviser.

At the heart of the TTC is that meaningful transformation through travel starts from within and leads to positive outcomes internally and externally. The aim is to inspire positive actions and outcomes from the individual, host and community, with a positive ripple effect. Members include DMOs, expedition cruises, safaris, tour operators specializing in wellbeing, mindfulness, adventure and nature, education/citizen science, spiritual, healing and wellness retreats. These trips can be for solo travellers, women only, families or blended travellers. Destinations which have joined include the Slovenian Tourism Board and Guyana Tourism.

> 'Travel is a privilege, and its full power is only realized when you release expectation, let go of entitlement, and engage for the purpose of fulfilment rather than entertainment.'
>
> **Jake Haupert, Co-Founder, Transformational Travel Council**

Summary

Already business leaders have seen glimpses of the future. The future could be zero travel or no more tourism if it fails to fundamentally transform and continues to cause environmental damages and social inequities.

The full force of the climate emergency is being increasingly experienced first-hand in tourism-dependent destinations around the world. Every day there are more reports of wildfires, flooding, droughts, heatwaves, biodiversity loss and mass climate migration. Warmer temperatures are leading to coastal erosion, coral bleaching and destroying the very locations and biodiversity that residents and visitors love to enjoy.

Travel and tourism demand in 2024 is breaking new ground in terms of peaks achieved. Constant growth is no longer viable – a new mindset of quality tourism is the only way forward. The new challenges faced by the sector from Generative AI to automation point to further uncertainty and constant adaptation to digital as well as sustainable transformation.

Everyone in the travel ecosystem from consumers, destinations and most travel brands appear to be enjoying the 'last hurrah': free to travel, aware that it might not be the right thing to do for the planet but carrying on regardless. There is growing awareness from consumers, and growing action from governments and travel businesses. Yet there needs to be a paradigm shift as called for by the UN. Destinations on the front line of the climate emergency like Fiji agree.

Unfortunately, the changes required to reach net zero in the travel industry are not going far enough. This means that targets will not be met – if ever at all. Many believe that failure on climate action could be the unfortunate outcome unless drastic and urgent steps are taken. A key challenge is how to transform such a disparate and fragmented sector so that every business, from micro to global, moves in the right direction on the path to long-term sustainability and profitability.

Endnotes

1 Euromonitor International (2024) from Euromonitor International/UN Tourism

2 Euromonitor International (2024) from Euromonitor International/UN Tourism

3 IMF (2024) www.imf.org/external/datamapper/profile/WEOWORLD (archived at https://perma.cc/SAE5-QA8G)

4 UN Tourism (2024) www.unwto.org/tourism-data/global-and-regional-tourism-performance (archived at https://perma.cc/9FNK-6JM9)

5 World Travel and Tourism Council (2024) wttc.org/research/economic-impact (archived at https://perma.cc/E8TL-4ZBP)

6 Ibid

7 UN Tourism (2023) www.unwto.org/news/women-take-centre-stage-in-tourism-development (archived at https://perma.cc/9R2Y-Z9XB)

8 Ibid

9 The Travel Foundation (2023) from WTTC and Lenzen et al., Envision 2030, www.thetravelfoundation.org.uk/wp-content/uploads/2023/02/Envision2030_SummaryFINAL.pdf (archived at https://perma.cc/M4V9-53HR)

10 IMF (2024) www.imf.org/external/datamapper/profile/WEOWORLD (archived at https://perma.cc/SAE5-QA8G)

11 Department of Business and Trade/Office of National Statistics, UK (2023) www.gov.uk/government/publications/hospitality-strategy-reopening-recovery-resilience (archived at https://perma.cc/UAK8-8VJU)

12 Financial Times (2021) Marriott warns of 'fight for talent' as hotels struggle to find staff (ft.com)

13 Marriott (2024) www.marriott.com (archived at https://perma.cc/K2GP-SP7L)

14 Ibid

15 Ibid

16 Ibid

17 NASA (2024) www.nasa.gov/news-release/nasa-analysis-confirms-2023-as-warmest-year-on-record/#:~:text=July%20was%20the%20hottest%20month,when%20modern%20record%2Dkeeping%20began (archived at https://perma.cc/N372-KLGR).

18 ECMWF as part of The Copernicus Programme (2024) https://climate.copernicus.eu/copernicus-june-2024-marks-12th-month-global-temperature-reaching-15degc-above-pre-industrial (archived at https://perma.cc/BSN6-YAJC)

19 Arctic Council (2024) arctic-council.org/explore/topics/climate/wildland-fire/ (archived at https://perma.cc/AGL4-D5FR)

20 IPCC (2022) www.ipcc.ch/report/ar6/wg2/chapter/chapter-15/ (archived at https://perma.cc/G7KF-HD6B)

21 UNEP (2020) www.unep.org/unepmap/resources/factsheets/climate-change (archived at https://perma.cc/TU7D-BX7J)

22 WWF (2024) www.wwf.org.uk/coral-reefs-and-climate-change (archived at https://perma.cc/XF4B-QSC3)

23 World Bank (2022) thedocs.worldbank.org/en/doc/c6aceb75bed03729ef4ff9404dd7f125-0500012021/related/mpo-fji.pdf (archived at https://perma.cc/LK2Q-SBLV)

24 Fiji Ministry of Finance (2024) www.finance.gov.fj/wp-content/uploads/2024/02/Fact-Sheet-Tourism.pdf (archived at https://perma.cc/2VEF-SHFP)

25 Tourism Fiji (2024) corporate.fiji.travel/statistics-and-insights (archived at https://perma.cc/HQX8-97GY)

26 Fiji Airways (2023) www.fijiairways.com/en-us/media-centre/island-of-vatulele/ (archived at https://perma.cc/4RHG-PWM8)

27 VisitValencia (2020) www.visitvalencia.com/en/news-room/valencia-becomes-first-city-world-verify-and-certify-carbon-footprint-its-tourist (archived at https://perma.cc/2M2K-FJEY)

28 Intrepid Travel (2024) www.intrepidtravel.com/uk/peru/machu-picchu/ do-you-need-a-permit (archived at https://perma.cc/HSR9-XZR4)

29 M Epler Wood, M Milstein and K Ahamed-Broadhurst (2019) Destinations at Risk: The invisible burden of tourism. The Travel Foundation, www. thetravelfoundation.org.uk (archived at https://perma.cc/M4V9-53HR)

30 Reuters (2023) www.reuters.com/world/asia-pacific/bhutan-cuts-daily-tourist-fee-by-half-lure-more-visitors-2023-08-26/ (archived at https://perma.cc/67PA-MGQM)

31 UN Global Compact (2024) unglobalcompact.org/what-is-gc/participants/search? button=&page=2&search%5Bkeywords%5D=&search%5Bper_page% 5D=50&search%5Breporting_status%5D%5B%5D=active&search%5Bsectors %5D%5B%5D=47&search%5Bsort_direction%5D=asc&search%5Bsort_ field%5D= (archived at https://perma.cc/8EGG-SUMM)

32 Science Based Targets Initiative (2024) sciencebasedtargets.org/ (archived at https://perma.cc/3H4J-G356)

33 Booking.com (2023) www.booking.com/articles/sustainable-travel-habits.en-gb. html (archived at https://perma.cc/6ABH-EHBA)

34 GSTC (2024) www.gstcouncil.org/wp-content/uploads/GSTC-Destination-Re-Assessment-Dubrovnik-2023.pdf (archived at https://perma.cc/B8RZ-J3RZ)

35 Eurostat (2024) ec.europa.eu/eurostat/web/european-statistical-system/w/ record-number-of-overnight-stays-in-accommodation-in-the-netherlands-statistics-netherlands-news-release (archived at https://perma.cc/4ATY-NJUC)

36 Ministry of Business, Innovation and Employment (2024) teic.mbie.govt.nz/ (archived at https://perma.cc/LRZ8-KNFC)

37 New Zealand Tourism (2024) www.tiakinewzealand.com/en_NZ/about-tiaki/ (archived at https://perma.cc/K642-G5AC)

38 4VI (2023) forvi.ca/wp-content/uploads/2023/04/4VI-ClimateActionPlan-April20.pdf (archived at https://perma.cc/66AL-AP22)

39 4VI (2023) 4VI Climate Action Plan, Synergy Enterprises

02

Balancing climate responsibilities with changing consumer values

Introduction

Increasingly, governments, businesses and consumers will face tough choices regarding the trade-off between collective versus personal responsibilities as the SDG milestones of 2030 and 2050 near. The growing need for urgent collective climate action will influence what consumers buy, who they buy from, how and where they travel. The most fundamental question will remain why they travel, if allowed to do so and what will be the cost benefit.

With each milestone will come a greater sense of urgency, awareness and ultimately despair as targets are unmet. The fight for climate justice and action, inspired by climate activists such as Greta Thunberg, is well underway. Yet green fatigue is setting in as consumers struggle with how to make the better, more planet-friendly choice and look to their favourite brands for help. Global brands are adapting their products and services to be more sustainable to meet new mandatory sustainability reporting as well as the UN SDG agenda. Fashion brands are embracing the circular economy and rejecting fast fashion. Car manufacturers are introducing electric vehicles, while food and beverage companies are removing single use plastic or expanding vegan options in fast food and restaurant chains.

A shift away from excessive consumption and extractive behaviours is required, as resources become ever scarcer as the global population expands. In tandem, the climate emergency demands ever smarter ways to adapt to and mitigate negative impacts from global warming, extreme weather events, natural disasters and impacts on local communities. Mass migration will lead to hundreds of millions of displaced climate refugees. Growing inequality caused by climate change and digitalization will further fuel tensions in transforming societies and businesses to a net zero future. The road ahead will be bumpy with many obstacles along the way, many yet unknown.

At the heart of travel and tourism lies a complex dichotomy. Hailed as a 'force for good' for its economic and social benefits, the travel industry is nevertheless a major contributor to climate change causing 8.1 per cent of manmade emissions pre-pandemic, especially transport accounting for almost 50 per cent.[1] Private and public sectors are uniting to face up to their joint responsibilities to decarbonize travel. The task to transition away from growth is Herculean, not for the fainthearted. A new path is required to strike the right balance, starting with a change in mindset in governments, businesses, local communities and consumers. A more mindful, conscious and inclusive type of travel will follow.

Global goals for collective action

The UN's 2030 Agenda for Sustainable Development sets the framework for how countries around the world are working together to reduce poverty, sustain peace and protect the environment. The agenda centres on 17 SDGs and their associated 169 targets that all 191 Members of the UN have signed up to. The SDGs were inaugurated in 2015, replacing the Millennium Development Goals.

Despite progress towards goals such as ending poverty, the pandemic temporarily reversed progress, which means even harder efforts are required to get back on track. The UN predicts that by 2030, 670 million people will still be suffering from extreme poverty, amounting to 8.4 per cent of the global population, only reducing poverty by 30 per cent compared to 2015.[2]

Even though travel and tourism are directly named in some goals as a vehicle to achieve them (namely 8, 12 and 14), the travel industry intersects with all goals. Travel businesses therefore play a significantly important role in achieving transformation across the key pillars of economic, social and environmental sustainability. Central to sustainable tourism are the Paris Agreement and SDGs, with a long-term goal of achieving net zero emissions by 2050.

Gearing up for 2030 climate targets

The SDGs are aligned with the Paris Agreement on climate change, implemented in 2016, to ensure that the global temperature does not rise over 1.5°c compared to pre-industrial levels. This entails signatory countries, of which there are 195 parties including the EU, promising to reduce their emissions and work to limit the impacts of climate change through adaptation and mitigation. Emerging countries receive climate finance to fight climate change and build resilience. Climate action plans are laid out in the form of Nationally Determined Contributions (NDCs). Every five years, climate action plans and targets are reviewed and ramped up to deliver the necessary transformations.

For 1.5°C to be achieved, the United Nations Framework Convention for Climate Change (UNFCCC) states that emissions must peak by 2025 and then fall by 43 per cent by 2030.[3] Some countries are moving faster than others. The EU has enshrined in law, the European Climate Law, its ambitious aim to reduce its greenhouse gas (GHG) emissions by 55 per cent by 2030, to be 'Fit for 55'. Even within the EU itself, there are fast movers such as Sweden that aims to reach carbon neutrality by 2045. It is not just advanced markets moving fast. Laos has set an ambitious target to reduce its carbon emissions by 60 per cent by 2030.[4]

The targets and frameworks are increasingly less abstract as laws and regulations come into play to guide the transition, especially in Europe with mandatory reporting in the form of the Corporate Sustainability Reporting Directive (CSRD).

Success is not guaranteed, but failure is not an option

The 2030 agenda is only the tip of the iceberg. The path to achieving net zero emissions by 2050 requires even greater action and commitment across all sectors. It is one of the biggest challenges that humanity faces. Already there are major concerns from the climate science and activist communities that the world is way off target.

The first ever global stocktake from the UNFCCC in 2023 revealed at COP28 in Dubai that there are major gaps in achieving the agreed GHG emissions targets, where emissions need to be reduced by 43 per cent by 2030 to meet the 1.5°C pathway. Despite more pledges to help finance transition in emerging countries, still trillions of dollars are required to guarantee success.

Climate science points out that the world is far from achieving success. UNEP's Emissions Gap Report in 2023 states that even with the current NDCs in place, the world is set for between 2.5°C and 2.9°C warming by next century.[5] Already scientists warn of catastrophic impacts when these temperatures are breached and go beyond thresholds where there is no way back, instigating irreversible changes on the planet. Near-term global tipping points are at risk of being breached in the North Atlantic, Greenland, West Antarctic ice sheets, coral reefs and permafrost regions. Positive actions such as decarbonization of transport and energy along with societal and behavioural changes can help to counteract negative impacts. Nature-based solutions are another critical way of helping to prevent tipping points, hence the importance of nature-based travel and tourism solutions.

REAL-WORLD EXAMPLE

Europe's Rail delivering sustainability for an integrated multi-modal transport system

Decarbonizing transport is the biggest challenge to face governments in the fight against climate change. Even countries like Germany that have a highly sophisticated integrated transport system have a long way to go before they fully electrify their railways with electrification currently at 54 per cent.[6]

Rail is earmarked in the European Union as a key way to deliver on sustainable passenger transport and freight through its Sustainable and Smart Mobility Strategy (SSMS). Governments cannot act alone. The EU has formed the Shift2Rail joint undertaking including rail manufacturers like Alstrom, Hitachi Rail, Bombardier, Network Rail, Thales, Siemens among other partners like Amadeus, Deutsche Bahn and SNCF.

Rail is highlighted as a means to solve problems like congestion, security of energy supply and climate change. The aim is to reduce carbon emissions by half for railway transport, double capacity carried while improving reliability and punctuality by 50 per cent.[7] However, the sector needs to transform and innovate to ensure it is a competitive, reliable and comfortable way to travel. The aim is to provide Mobility as a Service (MaaS) where multi-modes are available to passengers for a digital, inter-operable and seamless experience to ensure a door-to-door service on demand.

To deliver the EU-Rail objectives, it requires a multiple stakeholder approach across the spectrum as part of the Single European Rail Area (SERA), straddling the green and digital transformations required to meet the EU's Fit for 55 targets. Key to success is delivering a personalized and real-time service, that requires considering travel shopping, ticketing and trip tracking to improve the traveller experience. Such digital services are enabled by big data, GPS, automation, AI and IoT for smart mobility including intermediaries as well as transport providers.

Embracing innovative solutions for travellers and driving operational efficiencies to meet carbon targets will help to position rail in Europe, especially cross-border high-speed rail, as a viable alternative to other modes of transport like air. The forecasts for high-speed rail are for passenger traffic to double by 2030 and triple by 2050.[8] The price is a potential stumbling block especially when competing against budget airlines.

Yet with the revival of the night train in Europe, consumers are opting for a more sustainable way of travelling cross-border, harking back to a golden age of travel. Sleeper services are being brought back, connecting Paris to Berlin, or Vienna to Venice, for example with the reinstated Nightjet trains operated by ÖBB (Austria). New start-ups like European Sleeper, a cooperative, are launching new open access services, offering fun and smart travel, opening the market up to younger consumers. European Sleeper has so far raised EUR2 million from its community of night train fans.[9] For the luxury traveller, there is always Belmond's Orient Express following in the footsteps of Agatha Christie.

Moves to accelerate the shift to rail are undermined by supply challenges such as delays, cancellations, strikes and generally unhappy passengers.

TIPS

- To help close the sustainability 'say-do' gap regarding what consumers intend to do and their actual behaviour, more investment is required in new product development, transparency of impacts and marketing the benefits.

- There is no point reinventing the wheel when existing assets like rolling stock are already available and can be repurposed in a greener way.

- Greater efforts are required to open up previously government-controlled areas such as rail to start-ups.

- Even if the price is right and it ticks climate target boxes, if the customer experience is poor then uptake will be limited.

Accelerated path to net zero emissions by 2050

The long-term horizon to 2050 and beyond requires even more accelerated change. Already the 2050 agenda is laid out: to reach carbon neutrality and ensure net zero emissions. This entails ensuring that the 1.5°C pathway is achieved, and there are additional efforts to remove and store carbon using new technology such as carbon capture. The UN has called for a complete transformation to ensure that the world moves to net zero. With travel and tourism contributing to production, consumption and the movement of people, creating millions of tons of carbon every year, there is no time for complacency.

The WTTC's Roadmap to Net Zero outlines the share of GHG emissions by category, where transport accounts for 49 per cent, pre-pandemic, of which 17 per cent are derived from aviation (see Figure 2.1).[10] Rail stands out for its low carbon emissions yet its share of transportation itself remains under capacity, even in regions such as Europe with highly advanced rail connectivity. Spain's national rail operator, RENFE, even offered free tickets on its commuter and mid-distance journeys of less than 300 km in a bid to boost sustainable travel and lower emissions and provide relief to millions of travellers struggling with the cost of living crisis.

FIGURE 2.1 World GHG emissions from travel and tourism by category per cent share

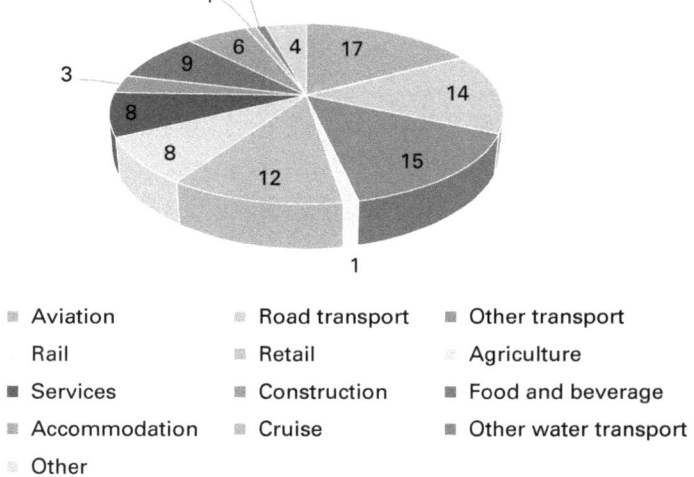

Aviation   Road transport   Other transport

Rail   Retail   Agriculture

Services   Construction   Food and beverage

Accommodation   Cruise   Other water transport

Other

SOURCE World Travel and Tourism Council 'A Net Zero Roadmap for Travel and Tourism' 2021
NOTE Pre-pandemic per cent share of world GHG emissions

Travel joins the race to net zero

The race to net zero emissions is officially on with the UN's initiative 'Race to Zero' including travel and aviation. Transport emissions rose 60 per cent over 2005–2015 and are forecast to increase by a further 25 per cent by 2030, highlighting the need for action.[11] Collaborative initiatives in travel such as the Glasgow Declaration for Climate Action in Tourism outline the pathway for transformation.

The Glasgow Declaration was formed by UN Tourism, The Travel Foundation, Tourism Declares and One Planet Network to join forces, launched at COP26 in Glasgow. The private sector is well represented in terms of pledges to produce and deliver on a climate action plan, being five times larger than destinations who have pledged in August 2024.[12]

Key sectors in travel have set their own net zero emissions targets. Aviation through the International Civil Aviation Organization (ICAO), International Air Transport Association (IATA) and their Carbon Offsetting and Reduction Scheme for International Aviation (CORSIA) aims for net zero emissions by 2050. Airports Council International (ACI) shares the same goal along with Cruise Lines International Association (CLIA) that is following the Getting to Zero Coalition's Call to Action for Decarbonization of Shipping. Meanwhile the Sustainability Hospitality Alliance (SHA) supports its members in reaching their net zero goals.

However, the first ever stocktake on tourism emissions by the Tourism Panel for Climate Change (TPCC) paints a sobering picture of the challenges and the scale of the transformation required. Sadly not one example of a destination or country exists for achieving the required reductions in emissions. In terms of TPCC respondents and their agreement with the potential success in meeting climate targets, 70 per cent believed that net zero targets for tourism would not be met by 2050, while an even higher percentage of 80 per cent agreed that aviation would not meet its net zero targets by then (see Figure 2.2).[13]

'At present, no country, no destination, and no subsector have achieved meaningful reductions in tourism greenhouse gas emissions. Overall, observed action and incremental change is insufficient to achieve the climate goals articulated in the Glasgow Declaration.'[14]

Tourism Panel For Climate Change

FIGURE 2.2 TPCC Stocktake Report on Tourism

Select Key Findings of Stocktake Report on Tourism

93% agreed that national tourism policy was not aligned with climate change in most countries

86% agreed that national climate policy does not integrate tourism sufficiently so the sector can contribute to NDCs

96% agreed that tourism policy does not support achieving national emission reduction targets

88% agreed that there was insufficient integration of emission reduction into destination management plans

88% agreed that there was insufficient climate adaptation into destination management plans

70%, 81% and 70% agreed that tourism, aviation and cruise industry would not achieve net zero targets by 2050 respectively

72% agreed that the impacts of climate change are already influencing tourism demand patterns

68% agreed that future climate impacts will restrict tourism development in many destinations by 2050 as climate change accelerates

SOURCE Tourism Panel for Climate Change[15]
NOTE Survey of 60 expert respondents in the field of climate change science and tourism from 31 countries

Travel businesses have their work cut out, as so far actions are not stretching far enough to make the necessary changes required to contain and adapt to climate change and mitigate negative impacts. There are clearly disconnects in terms of aligning country-level targets with tourism emissions where tourism is failing to be integrated into key government policies on climate change. Aviation sits outside countries' Nationally Determined Contributions (NDC) emissions commitments further complicating matters. This is a contentious issue that many are calling to be rectified.

Already the effects of climate change are felt in destinations around the world and will continue to intensify if global warming is not contained. These impacts will further determine future tourism demand patterns as popular destinations such as in the Mediterranean face ever higher temperatures. In turn, demand will shift to cooler climates or to different times of the year, reshaping seasonality which is no bad thing. Attention is already turning to new summer destinations from Scandinavia to the Baltics.

> TIPS
>
> • Different types of travel businesses must not view their emissions in silo.
>
> • Travel and tourism have a complex supply chain where scope 1, 2 and 3 emissions should be considered including partners and consumers.
>
> • Being honest about efforts to adapt travel businesses; sharing small successes will engage consumers and encourage participation.

Decarbonizing aviation fundamental to success

Tackling the elephant in the room

On the one hand, global aviation is one of the biggest success stories of the 20th century, thanks to innovations in jet fuel engines, technology and cheap fuel. Travel was democratized by the rise of the low-cost carriers (LCCs) in the late 1990s, such as Jet Blue (US), Ryanair (Europe) and AirAsia.

With their digital-first mindset, they stripped back services to the bare minimum to drive ancillary revenues from speedy boarding, baggage and seat selection. LCCs disrupted aviation and led the new era of budget travel. They moved out of hub and spoke airports into secondary airports, often point to point, opening up new destinations to the masses. The effects of this new wave of travel are still felt in destinations that suffered as carrying capacity was exceeded and air connectivity to some cities like Barcelona went out of control.

Yet, with the surge in air travel came a rise in emissions, accounting for 2 per cent of global GHG emissions, outpacing other modes of transport.[16] The sector is also one of the most challenging to decarbonize and abate, due to its high fossil fuel dependency, where alternative biofuels and sustainable aviation fuel (SAF) are only a mere fraction of total supply at 0.3 per cent in 2024.

Aviation is back and booming. According to IATA, 4 billion air passengers were carried pre-pandemic, with potentially 8 billion by 2040, doubling in size growing at 3.4 per cent compound annual growth rate (CAGR) (see Figure 2.3).[17] Each flight entails emissions throughout the entire journey such as taxing and fuel burn where the World Economic Forum (WEF) equates each air passenger-km with emissions of $83gCO_2e$.[18] There is clearly a huge disconnect between driving continuous demand and the need to decarbonize to hit climate targets.

FIGURE 2.3 Forecast air passenger traffic worldwide 2019–2050

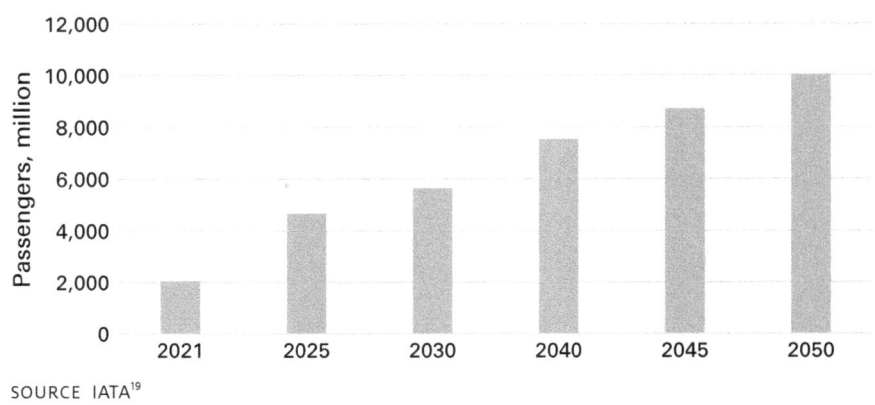

SOURCE IATA[19]

Aircraft manufacturers play an important role in delivering
a greener future

In terms of new aircraft deliveries over 2023–2042, Airbus forecasts that there will be demand for 40,850 new aircraft worldwide, with China and Asia Pacific accounting for almost half (46 per cent) of the deliveries.[20] Equally, aircraft renewals will be driven by the need to embrace modernization and upgrade to the next generation of aircraft that can deliver tangible fuel reductions to help reduce emissions. Already the next generation of Airbus aircraft are 25 per cent more fuel efficient.

Boeing and Airbus both point to the growth in single aisle aircraft which are the preferred type by LCCs. Boeing forecasts that by 2042, LCCs are expected to account for 41 per cent of the global single aisle fleet, amounting to 14,000 of a total 34,000 airplanes.[21]

There are other key priorities for Airbus such as innovating in aerodynamics, making the transition from 50 per cent to 100 per cent SAF, aligning with CORSIA targets, offsetting and carbon capture. The manufacturer is also exploring biomimicry to help reduce emissions and drive efficiencies. Taking inspiration from migrating geese flying in a V-formation, it has created its fello'fly concept with partners that entails two or more aircraft flying together to save energy during a long-haul flight.

In addition, Boeing's long-term forecasts point to faster growth in air traffic of 3.7 per cent CAGR 2023–2042 vs fleet growth at 3.2 per cent CAGR where the global fleet is forecast to almost double to 48,600 by 2042.[22]

In response to the pandemic, many airlines reconfigured their seat classes to dedicate to higher spending passengers. This has helped airlines to recover financially from the pandemic more quickly, but it entails a higher carbon footprint due to more seat space dedicated to premium passengers. A study by the World Bank reported that business class is three times higher the carbon footprint of economy class, while first class is nine times higher.[23] This short-term view of ramping up premium seats will not help to reach agreed targets.

ICAO provides a carbon emissions calculator (ICEC) to help aviation to measure their impact and help with offsetting. The calculator takes into account metrics such as number of passengers carried, passenger and cargo load factors, aircraft fuel burn and the distance flown along with seat class. Importantly, premium seat class produces twice the CO2e output than economy class for flights over 3,000 km.[24] The ICEC is available via application programming interface (API) for integration into business websites and mobile apps.

However, Professor Xavier Font, Professor of Sustainability Marketing, of Surrey University advised caution when using the ICAO carbon emissions calculator as 'it underrepresents aviation carbon emissions and that Google's carbon model would become the norm for understanding carbon impacts for aviation.' Google's Travel Impact Model (TIM) looks to predict future GHG emissions of flights based on origin-destination and aircraft type, also available via API.

Air passenger growth is expected to be particularly high in India and China, driven by strong economies and the rising middle class. It will be difficult to curb growth at the consumer demand level, so the faster the transition to next-generation aircraft and low-emissions fuels, the better.

TIPS

- Aviation should form part of a broader multi-modal system.
- Reconfiguring seat class away from premium (first class and business class) will help to reduce the carbon footprint of flying.
- Frequent flyer programmes (FFPs) need to be reinvented to promote sustainable travel behaviour and greener alternatives, beyond promoting more flights.
- FFP should address the high carbon footprint of members.
- In future, FFPs may be banned if deemed to perpetuate unsustainable travel.

- Working with companies that provide carbon footprint calculators or lifecycle assessments will help to drive sustainable operations and change consumer behaviour.
- Greener alternatives to flying should be offered to consumers.
- Transparency is critical to ensure the mainstreaming of lower carbon transport by helping travellers to understand their impacts.

REAL-WORLD EXAMPLE

easyJet explores carbon capture and next-generation aircraft with partners

easyJet is one of Europe's most popular LCCs carrying 82.8 million passengers a year.[25] Along with its partners, it is leading the pack when it comes to innovation, from trialling hydrogen fuel cell technology to looking at direct carbon capture to achieve its ambitions of reaching net zero emissions by 2050, if not before.

It has signed up to the Race to Net Zero initiative and is one of the first to work with the SBTi to produce its net zero roadmap, aiming to reduce emissions by 35 per cent by 2034, and reduce carbon emissions by 78 per cent per passenger by 2050.[26] The transition to SAF will play a pivotal role. SAF are made from biofuels from animal or plant material such as feedstocks or are synthetic, meeting sustainability criteria in terms of carbon emissions.

On the books, easyJet has 158 new Airbus NEO aircraft on order to help it make the transition to a more fuel-efficient and quieter fleet that are 15 per cent more fuel efficient and quieter.[27] It is equally investing in driving sustainability gains through its operational technology.

In a world first, easyJet ran tests on the ground with Rolls Royce, its engine manufacturer, testing green hydrogen from wind and tidal. The next step will be to run in-flight tests. The two companies are part of the UK's Hydrogen in Aviation (HIA) alliance, with the goal to deliver zero carbon aviation. Research conducted by HIA reveals that UK consumers support the need for moving to hydrogen to decarbonize aviation, where 71 per cent would be excited to fly on a hydrogen-powered plane.[28]

With easyJet's aircraft manufacturer partner, Airbus, looking to roll out hydrogen aircraft by 2035, the long-term outlook for decarbonizing aviation looks brighter. Its ZEROe programme is the means by which it will deliver the infrastructure, technology as well as the aircraft models to achieve its goals. The entire system needs to be transformed to accommodate the next generation of hydrogen fuel, working with airlines and airports such as in Norway, Sweden and Singapore.

easyJet and Airbus have also joined forces in terms of carbon capture where the former was the first airline to sign up to Airbus's new solution. Direct air carbon capture and storage (DACCS) entails taking carbon out of the atmosphere and storing it underground. Carbon capture is another tool on the road to net zero.

TIPS

- Partnership is critical to address the challenges of the road to net zero across the entire supply chain.

- Being transparent about carbon emissions with consumers, ways to offset and support the transition to SAF through incentives will be increasingly the norm.

- Demystifying the science and getting consumers excited about new technologies will help speed up adoption and push clean energy policies forward.

Regulatory framework and risks

International aviation falls outside the realm of the Paris Agreement and airlines are expected to work with ICAO to ensure they align with global climate targets. Central to achieving this is CORSIA which is the first global market-based measure. This scheme will be used in tandem with ongoing investment in new technologies and SAF.

ICAO announced its long-term aspirational goal (LTAG) to reach net zero by 2050 where there are no specifics given in terms of emissions reduction, instead it is left to the individual member states to determine their own targets and timelines. The LTAG confirms the sector's commitment to improving fuel efficiency by 2 per cent per annum, and keep net emissions at 2020 levels to be carbon neutral, and reach net zero by 2050.

CORSIA is the way that the sector will offset its emissions to ensure it is carbon neutral. The baseline for aviation emissions has been adjusted to be 85 per cent of 2019 levels, set for the period 2024–2034.[29] Until 2026, the offsetting scheme is voluntary only, and obligatory from 2027. Yet certain emerging countries such as SIDS are exempt.

Despite the best efforts by airlines and their suppliers, many travel leaders are questioning the future of air travel, considering whether the two should be decoupled considering that air travel is so highly carbon intensive. The path to net zero will not be smooth with pressing questions about whether aviation and tourism need to be broken up to deliver on agreed targets.

Greenwashing is a major concern. Major airlines including Delta Airlines, Lufthansa, Etihad, Ryanair, British Airways and KLM have been called out by regulators for overstating their environmental credentials or providing misleading claims. Greater transparency and accountability through verifiable green certification are called for, backed up by hard data. Only then can travel businesses claim to be providing a greener, better choice for consumers.

Thousands of lawsuits in the name of climate action are being filed against the aviation sector. Investor groups like Climate Action 100+ are applying pressure, pushing for capping flights to 2019 levels for business travel and long haul, and encouraging more rail travel to reach net zero. However, while there is a consumer backlash to greenwashing, some travel businesses are undertaking 'greenhushing', keeping quiet about their environmental and social impacts for fear of reprisal.

TIPS

- Tough decisions need to be made about the future role of aviation, where already travel leaders at the forefront of climate action are calling for aviation emissions to be decoupled from emissions from tourism.

- Aviation and airport services should be leading in terms of transparency, and look to transform into net-positive hubs of clean energy for the wider community.

- Avoid greenwashing at all costs.

- Do not engage in greenhushing as that is counter-productive to building awareness required to change behaviour.

- Build trust through transparency of reporting and certification, verified by data.

Transition to sustainable aviation fuel (SAF)

With transport and aviation accounting for 2 per cent of global emissions, and emissions growth expected to double in a business-as-usual scenario by 2050, IATA and ICAO are working to ensure that SAF is an integral part of a sustainable transition.[30] SAF can reduce CO_2 emissions by 80 per cent compared to jet engine fuel over its lifecycle, and requires limited modification to aircraft.[31] It can also be blended and dropped in with standard jet fuel offering a 40 per cent reduction in emissions.[32]

Airlines like Virgin Atlantic are leading the charge, where it conducted the first fully 100 per cent SAF-powered flight from London Heathrow to New York JFK Airport in 2023, known as Flight100. However, the airline was pulled up by the Advertising watchdog in the UK for potentially misleading consumers seeing as SAF-powered flights still emit emissions and are not 100 per cent sustainable.

To achieve the LTAG goal of decarbonization, 65 per cent of emissions reductions in aviation will be derived by the switch to SAF, 19 per cent from carbon offsetting and capture, 13 per cent from innovation such as electrification and hydrogen, and 3 per cent from more efficiencies in infrastructure and operations.[33] Yet the market for SAF is nascent, representing a mere fraction of global fuel supplies.

The other major stumbling block is the higher price of SAF due to its scarcity and lack of infrastructure, however, as the industry scales up the price will become more affordable. The way airlines go about encouraging consumers to pay for SAF is confusing and not directly aligned with the way they travel.

ICAO has announced its vision to reduce 5 per cent of aviation emissions through SAF and Lower Carbon Aviation Fuel (LCAF), where air travel is expected to produce 682 mtCO2e by 2030.[34] The majority of SAF supply and production will come through private and public sector collaboration and incentives. The rest will come through government-enforced production targets as seen in countries like the US aiming for 10 per cent by 2030. Airlines are also carving their own path and making SAF commitments. Yet the challenge ahead is immense, where 20 times more SAF is required to meet 2050 net zero targets.[35]

There are upsides to the conversion to SAF in terms of building resilient supply chains and energy security. Conversely, there are serious concerns that land cleared for feedstock to produce SAF destroys the environment and biodiversity. When decarbonization is a system-wide approach, removing nature-based solutions to make way for SAF production for aviation is in stark conflict with the global goals.[36]

NESTE POWERED BY BLOCKCHAIN TO DRIVE TRANSPARENCY

Companies such as Neste specialize in SAF fuel and are driving the transition to a circular economy. Neste partners with travel businesses following the SBTi and ensures that there is full transparency about the replacement of fossil fuel with SAF, which is verified via a third party and logged via the ISCC SAF credit registry. Its book and claim approach is powered by blockchain, enabling greater transparency and accountability. Delta is working with Neste to achieve its decarbonization goals, aiming for 10 per cent SAF by 2030.[37] Thai Airways trialled its first SAF-blended fuel flight using Neste MY Sustainable Aviation Fuel in 2023.

Blockchain book and claim solutions include the partnership between Amex, Shell and Accenture – Avelia – which is powered by Microsoft Azure and Energy Web's blockchain technology. Avelia is specifically designed to help airlines gain access to more affordable SAF and helps corporate companies to reduce the carbon emissions of their business travel activities that are a major contributor to companies' carbon footprints. Companies such as Google, Emirates and JetBlue have joined the scheme to purchase the environmental attributes associated with SAF.

The network of airports where SAF is being pumped into the system by Shell are Hong Kong, Changi (Singapore), Le Havre (France), Dubai (UAE), Ontario (Canada) and Los Angeles (USA).

However, any savings gained are not necessarily linked to the flight taken by the individual traveller. The book and claim system is therefore confusing for travellers. However, it has its benefits for corporate businesses as they can reduce their carbon emissions in one go through a single book and claim transaction and it is fully traceable for ESG reporting purposes.

End of business as usual

New pathways to explore

The pandemic taught the travel industry that change is constant and scenario planning is critical to build agility, drive resilience and adapt to changing threats. Whether it is contending with a wildfire, ash cloud, terrorism attack, natural disaster or any form of black swan event such as the Global IT glitch, risks need to be prepared for. When it comes to the climate emergency, there are different scenarios based on temperature rises versus pre-industrial levels. We do not fully know when tipping points will be passed and what types of climatic events will be unleashed. Greater awareness of the different pathways is essential to business planning.

Governments are also taking a critical view of the travel industry, as they look to ramp up their climate action. Led by the EU, countries like France, Germany, Spain and the Netherlands are taking concrete steps to stop excessive flying to help reduce carbon emissions. The French government in 2023 announced the ban of short-haul flights where an alternative rail journey of less than 2 hrs 30 mins is available. It also aimed to reduce the use of private jets. Spain is looking to emulate France's example and encourage uptake of its extensive high-speed rail network.

Germany is using taxation to discourage short-haul air travel with further increases in 2024. The Dutch government wanted to reduce capacity due to noise pollution and airport staffing but reversed its decision to cap flight capacity at Schiphol Airport, the country's hub. This led to opposition from the US government on behalf of JetBlue who said that it went against the US–EU Open Skies agreement. In 2024 Lufthansa announced its Environmental Cost Surcharge applied to departure flights from the EU, UK, Norway and Switzerland from 1 January 2025 due to the cost of paying for meeting climate commitments including CORSIA, SAF quotas and the European Trading Scheme (EU-ETS). The surcharge varies based on the route and fare type. This is in addition to Lufthansa's green fares offer where consumers can select sustainable options such as SAF and carbon offsetting at the time of booking with different options by seat class. Singapore announced the introduction of a green fuel levy on travellers to finance its SAF transition.

Government interventions and legal challenges to aviation expansion are just the beginning of a long road to curb emissions and achieve climate targets where tough decisions need to be made. As seen during the pandemic, governments can shut down the global travel industry, when required. If the travel industry fails to transform to a sustainable business model, it will find that it is increasingly at the mercy of regulation that it cannot control.

REAL-WORLD EXAMPLE
Envision 2030 outlines the challenges of balancing growth with decarbonization

The Travel Foundation's 'Envisioning Tourism in 2030 and Beyond' assesses the different pathways for travel and outlines the scale of transformation required.
Key findings include:

- more governments should include international aviation emissions in their Paris Agreement plans;

- tourism boards and travel companies targeting a greater proportion of short-haul customers and bringing net zero products to market;

- investing in greener forms of transport which are adopted and promoted;

- relying less on offsetting, focusing instead on decarbonization;

- fair policies that allow for differences in destinations around the world;

- slowing the expected rapid growth in aviation, with limits on the number of long-haul flights.

With a 'business as usual' scenario, travel and tourism are forecast to see carbon emissions double where carbon emissions could reach 4,500 million tons by 2100 which is not acceptable (see Figure 2.4).[38] Envision 2030 finds that there is only one pathway to achieve decarbonization and ensure global warming is kept at 1.5°C above pre-industrial levels. This pathway involves a system-wide approach to carbon emission tracking, in particular aviation. If change does not happen, the forecasts point to a future where long-haul flights will increase four-fold to account for 41 per cent of carbon emissions, but only account for 4 per cent of trips.[39]

The model used looked at different levers to achieve decarbonization balanced with industry growth, such as SAF production, carbon offsetting, taxation, traveller behaviour, infrastructural improvements, the shift to electric and travel speed. One of the most significant findings is to bring aviation into countries' NDCs plans to ensure a holistic view.

To avoid a climate crash by 2050, the tourism decarbonization scenario in Envision 2030 recommends:

- Mandate the use of SAF.

- Electrify lodging and transport, boost renewable energy.

- Drive funding of cleaner and more efficient modes of transport and technology from high-speed rail to EVs, including next-generation aircraft.

- Subsidize more sustainable ways of travelling.
- Limit aviation growth and cap long-haul flights to pre-pandemic levels.

Cruise is called out for having an 'extremely high carbon footprint' and to feature in NDCs. Another interesting way to achieve net zero travel includes disincentivizing frequent flyer programmes.

To ensure a fair transition, the report recommends that SIDs in the Pacific and Caribbean are provided with incentives to deliver change and diversify their economies. To drive resilience, there is a need to accept that tourism will never be the same and should not be.

'When it comes to the different pathways open to travel businesses and destinations, the options are becoming more limited as the target year of 2030 approaches. There are some signs of progress but it's inconsistent, with some moving faster than others. The bottom line is that the travel industry needs to accelerate decarbonization at scale and create positive impacts to ensure a fair and just transition, leaving no one behind.'

Jeremy Sampson, Chief Executive Officer, The Travel Foundation

---

TIPS

- Tourism futures research such as Envision 2030 shows that the current business-as-usual model is broken.
- Only system-wide transformation to a decarbonized, climate-positive tourism model is the way forward.
- Creative and radical approaches are needed to bring about the transformation required.
- Tough choices need to be made encouraging fewer long-haul trips, but longer trips along with trips closer to home.

FIGURE 2.4 Travel and tourism forecast emissions to 2100

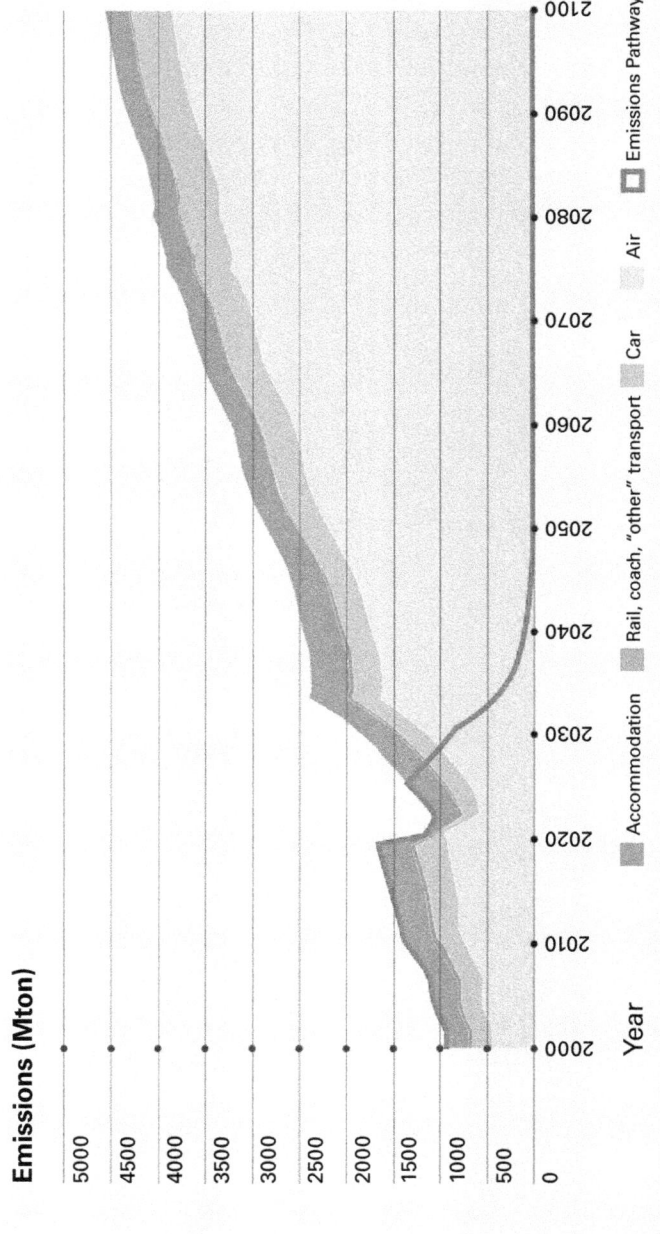

Emissions (Mton)

Year

■ Accommodation ■ Rail, coach, "other" transport ■ Car ■ Air □ Emissions Pathway

SOURCE The Travel Foundation, Envision 2030

Changing consumer values

Consumers care, but need help to make the right choice

Every year the warnings get starker about the irreversible effects of climate change as temperatures hit record highs. Consumers' climate awareness is at an all-time high and consumers are showing increasing climate consciousness year on year. However, there is a growing sense of fatigue setting in, so consumers need help to make the right choice for sustainable travel. This sense of helplessness is not helped by greenwashing claims by travel companies that destroy trust and make it hard for those companies that are delivering on their promises.

Booking.com's sustainability research reveals that sustainability is important but not a top priority when travelling (for 45 per cent of respondents), compared to 33 per cent that believe climate change is irreversible and their actions will not make any difference.[40] This sense of green fatigue and helplessness puts the onus on travel businesses and destinations to boost their sustainability actions to help empower consumers, where 75 per cent want to travel more sustainably and 71 per cent leave a place better than when they found it.[41] The 'say-do' behavioural gap must be reduced so that the travel positive choice is the easiest to pick. Promoting regenerative and positive impact travel and tourism clearly resonates with consumers.

Cost is an obstacle but can be overcome where 49 per cent believed that sustainable travel options are more expensive, and 49 per cent would appreciate being incentivized through rewards or discounts in making a more

FIGURE 2.5 Booking.com's Sustainable Travel Report key findings 2024

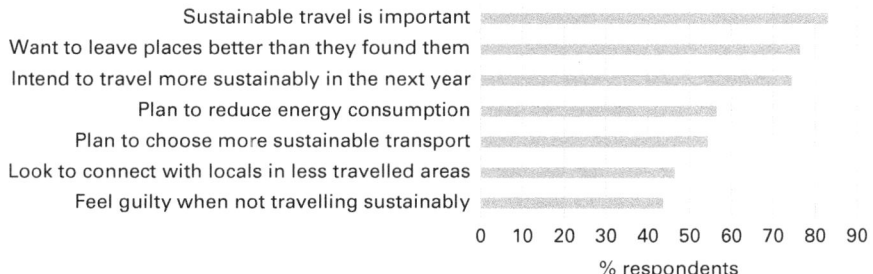

SOURCE Booking.com's Sustainable Travel Report (2024)[46]
NOTE 31,000 travellers in 34 countries

sustainable choice.[42] Behavioural science techniques are often used to encourage uptake of greener behaviours as well as gamification.

Trip.com's Sustainable Travel Consumer Report in 2022 supports that consumers are onboard where 79 per cent of consumers said that sustainable travel was vital.[43] In terms of who should shoulder the responsibility for sustainable travel: 33 per cent responded that it was the individual, 41 per cent for governments, and 37 per cent see it as a joint responsibility.[44] Booking.com's Sustainable Travel Report supports the collaborative approach to travelling sustainably with 44 per cent seeing travellers themselves as responsible, 43 per cent seeing it as the responsibility of the travel operators and 40 per cent looking to government to educate consumers on tourism impacts (see Figure 2.5).[45]

The multiple stakeholders involved in the travel supply chain and the interconnected nature of travel and tourism with adjacent industries and sectors make it a challenge for everyone to move in the right direction together.

Carbon offset controversy

In terms of carbon offsetting, there is a lack of awareness from consumers, where 80 per cent of passengers do not offset their flights, and 24 per cent do not want to offset their emissions.[47] This highlights the need for airlines to be more transparent and encourage uptake. Already 50 airlines are offering carbon offsetting programmes.[48] For example, Southwest loyalty members receive rewards when paying for offsets and the airline matches the payment. The EU Parliament ruling to ban misleading green claims and greenwashing highlights that trust begins with certification.

Controversy surrounds the carbon offset market where 'phantom carbon credits' have been found to be used to claim that flights are carbon neutral while no emissions are being offset. Such media exposure causes further consumer distrust in green claims and the entire system needs a reset.

Travel is a privilege, not a right

Consumer attitudes and lifestyles will continue to change in line with macro-economics, socio-demographics, megatrends and digitalization. It is important for travel businesses to ensure that the current and next generations of travellers are conscious that travelling is not a fundamental right but a privilege that comes with responsibilities. That responsibility starts with awareness of being a guest in another's home and a sense of steward-ship for the environment and biodiversity. Respect is fundamental to the relationship between guest and host community.

Gap Adventures founder Bruce Poon Tip in his documentary *The Last Tourist* explores the theme of how travel has become unsustainable and requires a change in mindset to view travel as privilege. Everyone has a role to play where travel businesses, governments and consumers all have a shared responsibility to travel in the most respectful way, and tread lightly. Educating travellers to make the better choice, and to be aware of the positive and nega-tive impacts is increasingly being front-loaded at the time of booking.

'Through Evaneos's direct to market model, at least 85% of revenue is spent in the destination. We provide more information and education at the time of booking, including impacts such as carbon footprint and fair wages. It is not effective to tell people what to do but rather educate travellers and nudge them to make the better choice.

We need to make bold choices to transform travel: This means giving up on over-crowded places, going off-season, creating alternative destinations and itineraries, off the beaten path which partner agencies are responding positively to.'

Marion Phillips, Head of Sustainability, Evaneos

The challenge is that people cannot always travel off-season or take more days off due to their employment contracts which leads to a new discussion with employers and governments to be part of the transformation to sustain-able travel.

Shift in strategic priorities for travel businesses

Positioning sustainability front and centre

Besides the hard-to-abate sector of aviation, travel businesses are uniting to find ways to work together to reduce their negative impacts. Collaboration takes many forms including global coalitions such as Travalyst, and members associations like the WTTC, SHA or GSTC.

Accommodation accounts for 6 per cent of global travel's carbon emis-sions (see Figure 2.1), increasingly lodging businesses are taking a holistic view of their scope 1, 2 and 3 emissions. Scope 3 emissions are the most challenging to tackle as these fall outside the direct control of the organization. To help large and small lodging businesses to transition to a zero carbon business model, useful resources are being designed such as The Hotel Sustainability Basics from the WTTC. Designed in partnership with Greenview, 'The Basics' are the minimum metrics for how to move to a more sustainable way of running hotels either as an owner or operator, aligned with the GSTC criteria.

The results of The Basics are shared in the Green Lodging Sustainability Report (GLSR) that encompasses metrics from 27,000 hotels including brands like Hilton, Jin Jiang, Marriott, Radisson and Rosewood. The GLSR outlines what are common, established, emerging to innovative practices across different pillars, showing that sustainability is an ever-evolving journey with different stages of evolution. Some common practices include the elimination of single use plastics such as straws (66 per cent of hotels) and nearly all (98 per cent) using a towel reuse programme, while offering recycling bins in hotels is less than a third (30 per cent of hotels) and generating renewable energy onsite has yet to take hold, only happening at less than a quarter of hotels.[49] Interesting innovation includes offering greater choices of plant-based menu options to guests: 16.7 per cent of all hotels offer vegan menu options for every meal versus 33 per cent offering vegetarian.[50]

THE HOTEL SUSTAINABILITY BASICS

Management and efficiency

1 measure and reduce energy use
2 measure and reduce water use
3 identify and reduce waste
4 measure and reduce carbon emissions

Planet

5 linen reuse programme

6 no single use plastic straws or stirrers

7 replace single use plastic water bottles

8 replace single use plastic mini toiletry bottles

9 green cleaning products

10 vegetarian options

People

11 community benefit

12 reduce inequalities

SOURCE   WTTC[51]

In hotels, technology is helping not just to drive operational efficiencies that reduce the bottom line but is a key enabler of meeting ESG targets. Amadeus research shows that 91 per cent of hotel businesses are using technology such as intelligent search to promote less-popular destinations to avoid overtourism, while 85 per cent use software to assess environmental impact and 76 per cent use Generative AI for sustainable travel option recommendations.[52]

There are leading global hotel chains that have been championing sustainability for decades, such as Iberostar and Radisson, working to keep global warming to 1.5C of pre-industrial levels. Both have committed to reducing scope 1, 2 and 3 emissions, while Iberostar is aiming to be carbon neutral by 2030, two decades ahead of many.

REAL-WORLD EXAMPLE

Radisson Hotels – long-term transformation to net-positive hospitality

Radisson Hotels Group is present in Europe, the Middle East and Africa (EMEA) and Asia Pacific, with a network of 1,250 properties in operation and development across 95 countries.[53] It was the first to introduce an environmental policy and has been focusing on being a sustainable and responsible business since 1989 through its core pillars of people, planet and community. It was acquired by Jin Jiang International, a leading hotel company in China, in 2019, and combined, it counts as one of the leading global hotel chains along with Marriott, Hilton and Accor.

Radisson is highly engaged with the travel industry and is a member of the SHA, WTTC, UN Tourism and Global Compact; it has signed the Glasgow Declaration and supported The Basics. Its achievements have set high standards for the sector. It was the first hotel chain to offer fully carbon-neutral meetings and events in 400 of its hotels, while 226 hotels are eco-label certified and 67 use 100 per cent renewable energy.[54]

As part of its SBTi target commitment, Radisson Hotels has a mid-term goal to reduce scope 1 and 2 emissions by 46 per cent by 2030, and a long-term goal to achieve net zero by 2050.[55] Central to achieving its net zero ambitions are the transformation to green buildings, green mobility initiatives such as EV charging and renewable energy. It reported over 700 electric car charging stations in 2022, with a quarter of its hotels using hybrid or fully electric taxis.[56] This integrated approach into the local multi-modal transport system is an interesting development.

To ensure delivery of change, it has an extensive network of Responsible Business teams across its operations, with climate champions. This is an under-developed area noted in the Green Lodging Report that travel businesses should focus on engaging climate champions. Its long-term partnership with water charity Just a Drop has helped fund 22 projects and deliver clean water and sanitation.

Scope 3 are among the most challenging emissions to tackle, where these are indirect and take place downstream along the value supply chain. Ways that Radisson Hotels aims to mitigate these is through adopting a circular economy approach, sustainable sourcing and procurement, and green mobility for its own operations and franchisees. Change is not just driven through operational strategy; through its loyalty programme, Radisson Rewards, guests can use their points to offset the carbon footprint of their stay.

Even if travel businesses do not own physical assets such as hotels, aircraft or fleet, there is still a major responsibility to tackle the environmental and social impacts of providing travel services. Tour operators, travel agents and online travel agents are increasingly taking a holistic view of their impacts and looking to account for scope 3 emissions that are outside of their direct control.

Companies that previously used their own sustainable labels have been removed in response to sustainability reporting requirements in the EU. For example, Booking.com removed its Travel Sustainable badges opting instead for third-party certification, following Dutch regulators stating that the programme was misleading to consumers. The Travalyst coalition's - which Booking is a member of - certification initiative aims to provide greater clarity and consistency in which certification schemes are used by accommodation providers in a fast-moving legislative space.

For tour operators like Intrepid Travel, G Adventures, Natural Habitat, The Travel Corp or Exodus Travel, it is part of their DNA to integrate sustainable practices across their operations and build positive impact travel experiences. Intrepid Travel is an industry pioneer in not just adventure travel but purpose-driven travel, one of the first major travel companies to become certified B Corp and adopt SBTi commitments.

Often adventure travel aligns well with sustainable travel. The Adventure Travel Trade Association (ATTA) has worked hard over the past two decades to ensure that their members are making the necessary transition to sustainable and regenerative tourism. In 2024, the ATTA announced a Sustainability Fund to help drive this transformation and invest in innovative solutions like Tomorrow's Air. A key component of the ATTA's work is to reduce revenue leakage and ensure that the maximum amount of tourism revenues stay in the local community which makes adventure travel so compelling to destinations.

REAL-WORLD EXAMPLE
Intrepid Travel – pioneering adventure travel company that leads by example

Intrepid Travel is an award-winning adventure travel company, entering its 35th year. It is performing well. In 2023, it reported travel bookings of AU$621 million, 330,000 customers in 2023 doubling revenue to AU$536 million on the previous year.[57]

Its focus on providing experiential and sustainable travel experiences for small groups resonates with consumers. Its destination management company (DMC) arm includes 27 different DMCs ensuring a positive impact on the ground. It operates in all regions of the world and offers tours in 114 countries. Its long-term goal is to become a AU$ billion dollar B Corp, while doubling the number of customers to 600,000 focusing on purpose rather than mere profit.

To ensure the company achieves its 2030 goals, it has three strategic priorities to drive excellence, adopt game-changing approaches and move into new lifestyles verticals such as lodging, media and adventure gear.

It is already making a move into lodging with the acquisition of Daintree Ecolodge in Australia in the Daintree Forest. Fundamental to growth is investing in its online presence, including providing more information about its tours through carbon labelling. Diversity, equity and inclusion (DEI) is extremely important where its Ethical Marketing Policy aims to ensure diverse and inclusive representation including Black, Indigenous and People of Colour (BIPOC), Lesbian, Gay, Bisexual, Transgender and Queer (LGBTQ), plus-size travellers, and Black and First Nations communities.

Intrepid Travel commits to following science-based targets by 2035, moving faster than its peers. However, it does not want to leave anyone behind in the travel industry and has shared its decarbonization guide as open source. It commits to reducing absolute scope 1 and 2 GHG emissions by 71 per cent by 2035 from a 2018 base year for its operations and trips business and reduce the carbon intensity from its trips emissions by 56 per cent per passenger day over the same period.

The ways to achieve these goals is a combination of reducing flights offered, new product development, SAF, lodging suppliers using renewables plus aiming for net zero emissions among other activities. Some products that have already been removed from itineraries include carbon-intensive activities such as helicopter rides, snowmobiling and scenic flights, while a switch to vegetarian meals would reduce emissions from meals by 50 per cent.

Social impact is another strong pillar for Intrepid, where it has launched initiatives with First Nations to drive inclusion, reconciliation and sales with an 8 per cent increase reported for First Nation businesses in 2023. The Intrepid Foundation raised AU$2.8 million for local communities and projects.

However, despite its rigorous approach to sustainability, there was a complaint in the UK from the Advertising Standards Authority (ASA) regarding an advert with the term 'planet-friendly' which was subsequently removed as misleading. The company does not offset carbon emissions from flights or lodging that its customers book independently to join the start or end of a trip, whereas all other emissions are accounted for through offsetting.

Summary

The race to net zero is on. It is a global cross-sector effort involving multiple stakeholders from government, industry, communities and consumers. Global frameworks are in place to deliver the required transitions and meet agreed targets. Decarbonization is complex. Eradicating poverty and ensuring a fair, just transition is not straight-forward, but all are critical if the planet is to continue to thrive for generations to come.

There are several pathways to deliver sustainable development, however, not all of them are open or even viable for travel. Governments are intervening increasingly to speed up the shift to a low-carbon future. Transport as a major carbon emitter is in their sights as seen with short-haul flight bans, taxation, and future caps on airport expansion and long-haul flights, just for starters.

Creative and bold approaches are required, adopting a test and learn mindset. If things work, the approach is to scale fast and ensure that no one is left behind. Already travel brands like Intrepid Travel are embracing open source and sharing best practice. Competitors from Expedia to Booking are coming together, led by a common purpose.

Yes, mistakes are being made. Greenwashing can happen even to the most planet-conscious brands, but this must not derail the level of commitment required to press on with ambitious science-based targets.

Every time a consumer is asked to reuse a hotel towel, forgo single use plastic, offered the choice of a vegetarian meal, to buy local or pledge to tread lightly, success becomes one small step closer. These small behavioural changes should be rewarded to accelerate progress. The consumer mindset is already changing yet it is hard to consistently make the right choice. Consumers are onboard with not just sustainable travel but regenerative travel, understanding the reasons why. They simply need help in making the better choice that ticks the experiential factor yet is mindful of impacts. Transparency through technology builds trust, where price becomes less of an obstacle to the uptake of sustainable travel.

Endnotes

1 WTTC (2022) wttc.org (archived at https://perma.cc/XV9F-GTEQ)

2 United Nations (2023) sdgs.un.org/goals/goal1#progress_and_info (archived at https://perma.cc/2T4Y-P2PR)

3 United Nations Framework Convention on Climate Change (2024) unfccc.int/process-and-meetings/the-paris-agreement (archived at https://perma.cc/ERC2-ZDY8)

4 UNDP (2023) climatepromise.undp.org/what-we-do/where-we-work/lao-pdr (archived at https://perma.cc/EGP9-AZWH)

5 UNEP (2023) www.unep.org/resources/emissions-gap-report-2023 (archived at https://perma.cc/32T4-LMAU)

6 European Commission (2022) alternative-fuels-observatory.ec.europa.eu/transport-mode/rail (archived at https://perma.cc/C6FL-ADX8)

7 Ibid

8 European Parliament (2024) www.europarl.europa.eu/RegData/etudes/BRIE/2023/754599/EPRS_BRI(2023)754599_EN.pdf (archived at https://perma.cc/V9SM-5RCK)

9 European Sleeper (2024) europeansleeper.eu/about-european-sleeper (archived at https://perma.cc/WL6Z-C6M3)

10 WTTC (2021) wttc.org/Portals/0/Documents/Reports/2021/WTTC_Net_Zero_Roadmap.pdf (archived at https://perma.cc/G4BE-BZJ8)

11 One Planet Network (2022) www.oneplanetnetwork.org/sites/default/files/2022-02/GlasgowDeclaration_EN_0.pdf (archived at https://perma.cc/RTR8-LT9S)

12 One Planet Network (2024) oneplanetnetwork.org/programmes/sustainable-tourism/glasgow-declaration/signatories?aggregated_field=&items_per_page=12&sort_by=created&sort_order=ASC&f%5B0%5D=filters_organisation_glasgow_signatory_type%3A3474 (archived at https://perma.cc/W2Q7-Z6VJ)

13 Tourism Panel for Climate Change (2023) Tourism and Climate Change Stocktake 2023 (Eds S Becken and D Scott, tpcc.info/stocktake-report/ (archived at https://perma.cc/38P2-BJFM)

14 Ibid

15 Ibid

16 International Energy Agency (nd) www.iea.org/energy-system/transport/aviation (archived at https://perma.cc/A5Y8-CE3K)

17 IATA (2023) www.iata.org/en/iata-repository/publications/economic-reports/global-outlook-for-air-transport---december-2023---report/ (archived at https://perma.cc/ZW4A-S4R2)

18 WEF (2023) www.weforum.org/publications/net-zero-industry-tracker-2023/in-full/aviation-industry-net-zero-tracker/ (archived at https://perma.cc/263A-4GHF)

19 IATA (2021) www.iata.org/en/programs/environment/flynetzero/ (archived at https://perma.cc/7ULF-U7X7)

20 Airbus (2023) www.airbus.com/en/products-services/commercial-aircraft/market/global-market-forecast (archived at https://perma.cc/9ETB-3HHX)

21 Boeing (2023) cmo.boeing.com/ (archived at https://perma.cc/WA9Y-649C)

22 Boeing (2023) investors.boeing.com/investors/news/press-release-details/2023/Boeing-Forecasts-Demand-for-42600-New-Commercial-Jets-Over-Next-20-Years/default.aspx (archived at https://perma.cc/G96E-2CNU)

23 World Bank (2013) documents1.worldbank.org/curated/en/141851468168853188/pdf/WPS6471.pdf (archived at https://perma.cc/7A3G-EUM5)

24 ICAO (2024) www.icao.int/environmental-protection/CarbonOffset/Pages/default.aspx (archived at https://perma.cc/6D3P-TBF5)

25 easyJet (2023) corporate.easyjet.com/investors/regulatory-news/news-details/2023/easyJet-PLC---Final-Results/default.aspx (archived at https://perma.cc/T8CA-8K5J)

26 easyJet (2024) www.easyjet.com/en/sustainability (archived at https://perma.cc/P55U-C5GY)

27 easyJet (2023) corporate.easyjet.com/investors/regulatory-news/news-details/2023/easyJet-PLC---Final-Results/default.aspx (archived at https://perma.cc/T8CA-8K5J)

28 easyJet (2023) mediacentre.easyjet.com/story/16167/uk-hydrogen-alliance-established-to-accelerate-zero-carbon-aviation-and-bring-an-34bn-annual-benefit-to-the-country (archived at https://perma.cc/GQ6M-N3S4)

29 IATA (2023) www.iata.org/en/iata-repository/pressroom/fact-sheets/fact-sheet---corsia/ (archived at https://perma.cc/425C-TBYK)

30 Transport and Environment (2024) www.transportenvironment.org/challenges/planes/airplane-pollution/ (archived at https://perma.cc/PMK2-8F9E)

31 Neste (2024) www.neste.com/ (archived at https://perma.cc/L4U5-VK5Q)

32 Avelia (2022) aveliasolutions.com/ (archived at https://perma.cc/7YC8-8CQA)

33 IATA (2023) www.iata.org/en/iata-repository/pressroom/fact-sheets/fact-sheet---alternative-fuels/ (archived at https://perma.cc/2THL-WUAU)

34 IATA (2023) www.iata.org/en/iata-repository/pressroom/presentations/saf-gmd2023/ (archived at https://perma.cc/9ZU9-KQW7)

35 Ibid

36 Susanne Becken, Brendan Mackey, David S Lee, Implications of preferential access to land and clean energy for Sustainable Aviation Fuels (2023) www.sciencedirect.com/science/article/pii/S0048969723025044 (archived at https://perma.cc/FVT2-32RP)

37 Delta (2024) news.delta.com/unlocking-potential-sustainable-aviation-fuel-through-meaningful-advocacy#:~:text=Delta%20has%20signed%20agreements%20with,by%20the%20end%20of%202030 (archived at https://perma.cc/KS45-M2LV)

38 The Travel Foundation (2023) P Peeters, B Papp, 2023. Envisioning Tourism in 2030 and Beyond. The changing shape of tourism in a decarbonising world, www.thetravelfoundation.org.uk/envision2030/ (archived at https://perma.cc/6DK4-NB4N)

39 Ibid

40 Booking.com (2024) news.booking.com/latest-bookingcom-sustainable-travel-data-reveals-ongoing-challenges-for-consumers--highlights-a-heightened-opportunity-for-cross-industry-collaboration/ (archived at https://perma.cc/UG2T-NTX5)

41 Ibid

42 Booking.com (2023) news.booking.com/download/31767dc7-3d6a-4108-9900-ab5d11e0a808/booking.com-sustainable-travel-report2023.pdf (archived at https://perma.cc/C9DV-UC9N)

43 Trip.com (2022) pages.trip.com/images/group-home/media/downloads/Sustainable%20travel%20consumer%20report.pdf (archived at https://perma.cc/NE5H-WQTF)

44 Trip.com (2022) pages.trip.com/images/group-home/media/downloads/Sustainable%20travel%20consumer%20report.pdf (archived at https://perma.cc/NE5H-WQTF)

45 Booking.com (2024) news.booking.com/latest-bookingcom-sustainable-travel-data-reveals-ongoing-challenges-for-consumers--highlights-a-heightened-opportunity-for-cross-industry-collaboration/ (archived at https://perma.cc/UG2T-NTX5)

46 Booking.com (2023) news.booking.com/download/31767dc7-3d6a-4108-9900-ab5d11e0a808/booking.com-sustainable-travel-report2023.pdf (archived at https://perma.cc/C9DV-UC9N)

47 IATA (2022) www.iata.org/contentassets/baf7cb5eed64472aaac8906608085aff/global-passenger-survey-2022-media-briefing.pdf (archived at https://perma.cc/96K5-UGA2)

48 IATA (2024) www.iata.org/en/programs/environment/carbon-offset/#:~:text=Over%2050%20airlines%20have%20introduced,a%20third%2Dparty%20offset%20provider (archived at https://perma.cc/6S78-NKJ2)

49 Greenview (2022) greenview.sg/wp-content/uploads/2022/12/Green_Lodging_Trends_Report_2022.pdf (archived at https://perma.cc/4BR3-VLKF)

50 Greenview (2022) greenview.sg/wp-content/uploads/2022/12/Green_Lodging_Trends_Report_2022.pdf (archived at https://perma.cc/4BR3-VLKF)

51 WTTC (2024) wttc.org/initiatives/hotel-sustainability-basics (archived at https://perma.cc/C98B-SEHL)

52 Amadeus (2024) www.amadeus-hospitality.com/2024-travel-tech-investment-for-hospitality/?submissionGuid=ddbdf3cc-f096-4252-88df-3822e3efcfac (archived at https://perma.cc/R6R5-WSJE)

53 Radisson (2024) www.radissonhotels.com/en-us/corporate/about-us/our-presence (archived at https://perma.cc/RQM9-7XMY)

54 Radisson Hotels Group (2022) media.radissonhotels.net/image/responsible-business--corporate-use-only/miscellaneous/16256-142211-m26438577.pdf (archived at https://perma.cc/9C7L-QB85)

55 Radisson Hotel Group (nd) https://www.radissonhotels.com/en-us/corporate (archived at https://perma.cc/8YYS-7JVN)

56 Ibid

57 Intrepid Travel (2023) reports. intrepidtravel.com/Intrepid-Integrated-Annual-Report-2023.pdf (archived at https://perma.cc/AW5A-LWKG)

03

Understanding future
traveller demand

Introduction

Future traveller demand will increasingly come from emerging markets as urbanization and income distribution power middle-class growth. Asia Pacific is at the forefront of these seismic shifts led by India and China. The Asian Century is upon us and next century is predicted to see Africa become a key source of demand for goods and services. Advanced regions like North America and Europe will continue to age, with Baby Boomers already replaced by Millennials and future generations like Alpha driving consumer trends. Future travellers will therefore be younger, more diverse and climate conscious than previous generations. They will identify as experience-seekers and digital natives seeking personalization.

Discretionary income will remain a prerequisite for travel, and travel will continue to hold strong appeal as consumers increasingly opt to prioritize experiences over physical goods. The appeal of new, authentic and local experiences will continue. People will still follow their passions, mixing them with travel.

With economic growth, a new generation of middle-class consumers will enter the marketplace for travel and tourism. Expanded air connectivity will drive demand for travel especially in emerging regions like Asia Pacific, while intra-regional travel will continue to hold the most appeal due to more affordable prices and no-fly or lower carbon alternatives.

Yet entry to the middle class may not necessarily translate directly into access to international travel. New bans, restrictions, higher taxation and stricter entry criteria, such as a longer length of stay or smaller carbon footprint, will be increasingly applied to potential visitors. Financial means will no longer be the only way to unlock travel experiences. It is important to ensure that the path to the decarbonization of travel remains inclusive, not exclusive.

Future travellers shaped by megatrends

Demographic shifts

Looking at the demographic outlook for the world over the next few decades, the forecast is for increased population growth albeit at a slower pace. It is expected that 10.4 billion people will live on earth by 2100, up from the current 8 billion and 9.7 billion by 2050.[1] With increased population growth will come further economic development driven primarily by the emerging regions of Africa, Middle East, Asia Pacific and Latin America. On the other hand, the advanced regions of Europe and North America will struggle with the challenges of an ageing population and lower birth rates.

In line with economic development, the middle class expands and creates more discretionary income to spend on experiences and goods. Already lower-middle and upper-middle income account for the bulk of the population, at 75 per cent.[2] The emerging and least-developed countries are expected to enjoy higher economic growth, driven by a complex range of factors from a younger population entering the workforce, technology, and private sector and public sector investments. International development organizations such as the World Bank are working to ensure that there is levelling up and a fairer and more equitable spread of wealth to eradicate extreme poverty and deep-rooted inequalities from a long history of colonization.

The long-term potential for sub-Saharan countries is enormous with the 22nd century Africa's for the taking. Africa's young population coupled with digitalization will help to forge a strong path ahead. Travel and tourism already play an important role in sustainable development, where countries like Botswana have been pioneering high-quality, low-volume tourism for decades.

Urbanization and megacities

In turn, urbanization accelerates as people gravitate to cities for employment and opportunities. The urban population already stands at 55 per cent, and is expected to reach almost 70 per cent by 2050 and continue to increase to 85 per cent by 2100.[3] Increased urbanization is a driver of economic growth and employment that in turn fuels consumer spending. Yet urban growth causes additional environmental and social stresses especially where social safety nets are not well established and weaken resilience. Ongoing urbanization puts further pressure on infrastructure, transport, mobility and ecosystems, requiring smart city strategies to transition to a net zero future. Cities increasingly find themselves at the forefront of climate change. Challenges are multi-fold from pollution, migration, displacement, water risks and food shortages that ultimately result in conflict.

The rise of megacities where residents exceed 10 million people will continue to spur economic growth where there are 33 megacities worldwide in 2024, with that number set to reach 50 by 2050.[4] The largest megacities reside in Asia and Africa, and those regions are also forecast to see the greatest growth in megacities by 2050: Mumbai (India), Delhi (India), Dhaka (Bangladesh), Kinshasa (Congo, DR) and Kolkata (India).[5] Where megacities thrive so do large capacity airports, so-called aerotropolis such as Singapore Changi Airport or Dubai International Airport. Not surprisingly, urban consumers fuel demand for travel and tourism where propensity to travel is tied to disposable income as air connectivity, infrastructure development and mobility investments follow urbanization.

As seen in countries like China, urban consumers enjoy visiting rural locations, whether to visit friends and relatives, or immerse themselves in nature and get a change of scene from their daily lives. The Golden Week holidays for Chinese New Year and National Day in October in China every year, see the mass movement of people from urban to rural locations. This is time for urban residents to reconnect with loved ones and visit popular destinations such as the Great Wall of China, Beijing, Shanghai, Chengdu or enjoying parts of the ancient Silk Road at Dunhuang.

REAL-WORLD EXAMPLE
India – huge potential as a high-spending source market, driven by digital and experience

By 2030, Indian international travellers will be one of the top largest source markets in the world, ranked in sixth place.[6] India is the world's largest populous nation and a major driver of the global economy along with China due to strong public investment. India is becoming richer thanks to public and private investment, where the middle class is forecast to double to 61 per cent of the population by the time the country reaches its centenary in 2047, accounting for over 1 billion people.[7] The long-term outlook for India bodes well for accelerated outbound spending on travel and tourism as the middle class grows and has the means to travel abroad, aided by a highly developed aviation market with strong domestic airlines. Domestic tourism will also gain a welcome boost with the country's excellent rail network and air connectivity.

India travel businesses like MakeMyTrip understand the Indian traveller, their aspirations and preferences, where 100 million Indian consumers use the platform yearly. Trends such as staycations, experiential travel experiences with an emotional connection with cultures and communities, and spiritual tourism experiences are popular. Spiritual destinations such as the ancient cities of Ujjain and Ayodhya are important sites for Hindu pilgrims and enjoy strong interest from Indian travellers. Popular short-haul destinations abroad include Dubai (UAE), Singapore and Bangkok (Thailand), Toronto (Canada), London (UK) and New York (US).[8] The historical links with the UK means that there is a strong visiting friends and relatives (VFR) market between the two countries, for family, education and work.

In terms of travel companions, travelling as a family saw the highest increase, followed by solo travel with May and December being the peak travel months.[9] This makes the Indian outbound market more attractive as they travel in off-season months seen in the Global North. Indian travellers enjoy travelling as a family group and that opens up opportunities for family rooms in hotels as well as short-term rentals with greater flexibility and space.

An important aspect of travelling abroad is the ability to pay, where most Indian consumers pay for travel by credit card and 36 per cent by Unified Payments Interface (UPI) which is a real-time payments system unique to India.[10] UPI is also accepted in select markets like the UK, France including popular visitor attractions like the Eiffel Tower, Singapore, Malaysia and UAE. For more countries to encourage Indian visitors, providing greater levels of personalization will be key including offering vegetarian options, no-alcohol alternatives and UPI acceptance to enable seamless payments. Another major consideration is travel facilitation, providing ease of entry cross-border such as visa-free travel, which is only offered to Indian citizens in under 30 countries. India holds enormous value potential and it would be unwise to ignore such a dynamic source market with long-term opportunities to support sustainable travel transformation.

'Emerging source markets like India will not take kindly to being preached at and told by the Global North that they cannot fly when the West has been flying for the past few decades. A more impactful approach to achieve sustainable travel behaviour would be to tap into the rich pro-environmental historical traditions of emerging markets like India and inspire the new generations of travellers.'

Dr Sumeetra Ramakrishnan, Associate Professor, School of Hospitality and Tourism Management, Centre for Competitiveness of the Visitor Economy, Gender, Entrepreneurship and Social Policy Institute, Surrey University

Income tipping point for international travel

The tipping point for travel is when consumers enter the middle class around US$10,000 per capita leading to greater propensity to fly.[11] The strong correlation between income and travel continues to boost demand where there is connectivity, visa facilitation and infrastructure in place. All these activities in turn drive carbon emissions, unless there is a change in attitude and approach. The Paris Agreement and 2050 net zero pathway require a decoupling of economic growth from fossil fuels consumption to enable decarbonization and meet climate targets.

Asian Century untaps new opportunities

The 21st century is hailed as the Asian Century, led by China and India, its two demographic and economic powerhouses, as well as the overall high-growth economies of ASEAN. By 2030, 3.5 billion affluent middle-class Asian consumers are expected to be spending in Asia and beyond.[12]

With rising incomes comes greater propensity to travel, especially if facilitated by strong air connectivity and an extensive network of low-cost carriers with a low entry price to international travel. The ASEAN Single Open Sky has been instrumental in liberalizing air transport intra-regionally with agreements also agreed with the EU in 2022 and China since 2012. Due to the geographical distances between Asian countries, flights tend to be mid haul, rather than short haul. This entails a larger carbon footprint. Already some of the most popular domestic flight routes such as Beijing to Guangzhou (China), or Beijing to Sanya in Hainan (China) are over three hours.

Asia Pacific as a region is the dominant focus for expansion across all types of global travel brands from hotels to online travel agents (OTAs), luxury goods players and equally of great interest to destinations across the world. Asia Pacific has been the focus of hotel pipeline development for the past few decades, especially China.

TIPS

- One size does not fit all for Asian travellers with important values and cultural differences to consider.

- Some Asian countries like China and Japan are ageing which leads to more inclusive and accessible travel services to meet the needs of older travellers.

- Digitalization across the customer journey is expected as standard by Asian travellers that are mobile-first consumers.

Geopolitics plays a role in who travels where

There are other factors such as macroeconomics and geopolitics to consider, such as the polarization in diplomatic relations between the world's largest economies, the US and China. These tensions spill over into the realm of travel and tourism where an increasingly fragmented political order complicates business operations, where travel businesses and tourism boards can be caught up in political tensions and conflict. Military conflicts, civil unrest and war can have a devastating toll on local communities and regions, taking years for countries to recover their tourism market and potential. Conflict in the Middle East region over the years has not deterred visitors to the region, with it being one of the first to recover from the break in international tourism after the pandemic.

In Europe, the aftermath of the war in Ukraine will also require huge investment of multi-billions to rebuild the country when peace is finally achieved. The transition to an inclusive and climate-resilient country, steeped in its own culture and history will enable Ukraine to open up to international tourism and become a destination of the future. Global travel brands are already on hand to help provide the necessary support in marketing the destination once safe.

There has been a sea change in the way that the rest of the world regards the advanced markets, with a Global North–Global South divide and East–West split. These tensions tend to flare up when it comes to funding to pay for loss and damage suffered by emerging markets as compensation for damages incurred from climate change. This rise in tensions may feed into the type of welcome and attitudes visitors may receive.

Middle East airlines from their global hubs – Emirates (Dubai), Qatar Airways (Doha), Etihad (Abu Dhabi) – will continue to benefit from air liberalization and their government ownership to connect the seven regions of the world. Turkish Airlines (Istanbul) which is part government owned has also enjoyed strong expansion. These airlines have had the Indian subcontinent and Africa firmly in their sights as the balance of power shifts away from the north to south.

Future migration considerations

Official tourism statistics exclude the movement of migrants, asylum seekers and refugees. However, the number of people leaving their homes or countries is significant and increasing, often using the same modes of travel and transport as domestic and international visitors. Mass displacement can be caused by multiple factors including conflict but more frequently by climate change and natural disasters.

The UN Refugee Agency (UNHCR) expects 130 million people to be displaced in 2024, with the majority (52 per cent) in war-torn countries of Syria, Afghanistan and Ukraine, seeking shelter in countries like Iran, Turkey and Germany.[13] By 2050, there could potentially be over 1.2 billion climate refugees, where most will gravitate to cities to look for work and opportunities.[14]

Countries in Asia Pacific are at most risk of climate change impacts such as flooding on low-lying coastal areas including China, India, Bangladesh, Indonesia, Vietnam and Thailand. Some of these most at-risk countries are also highly dependent on tourism, requiring not just a view of tourism demand but equally climate-related mobility patterns to build resilience.

Flows of remittances are another useful way of understanding migration and predicting future demand for travel and transport services, where the largest recipients of remittances are India, China, the Philippines, Mexico and Egypt.

TIPS

- To achieve truly sustainable tourism, it is important not to be derailed by short-term political goals when new governments take the reins, instead taking a long-term view.

- For travel to be a force for good, it must be apolitical with a responsible use of soft power.

- Informal travel patterns should be accounted for when assessing destination management and future demand predictions.

- Migration should be formalized into the tourism satellite accounting system so that it is more inclusive and promotes safe travel for all.

- More than 10 per cent of the world's population will be travelling informally by 2050 and are an untapped source of labour, opportunity and spending power.

New holistic approach for optimizing the value of source markets

Future tourism demand patterns will be shaped by converging factors such as global megatrends like urbanization, shifting economic power, new geopolitical realities and ever extreme climate impacts. The outlook to 2030 for outbound tourism spending is characterized by Asia Pacific overtaking Europe to be the largest spending region on international travel, led by China and India, the world's most populous countries. Consumers in regions like Asia Pacific are living day in, day out with the impacts of climate change. In Asia Pacific, 64 per cent of consumers believe climate change is an emergency and 56 per cent have already changed their behaviour or purchasing to respond to climate concerns.[15]

India is forecast to be in the top 10 markets for outbound travel, while China will usurp the US to be the largest spender (see Figure 3.1).[16] Advanced markets of the US, UK, France and Germany will continue to offer strong potential. However, as travel brands and destinations look to the long-term outlook, the value potential of source markets (or origin markets) will become over-shadowed by their comparative holistic values. A more critical view will be applied in terms of what defines value, going beyond geography, income, sex and psychographics. The criteria of high spending will be replaced by a more complex matrix of metrics across values, attitudes, behaviours and impacts: for example, length of stay, mode of transport, dietary preferences such as meat-free, use of public transport and climate-friendly mobility from bike-share to EVs. In the same way that organizations

FIGURE 3.1 Forecast outbound tourism spending by key source markets 2024/2029

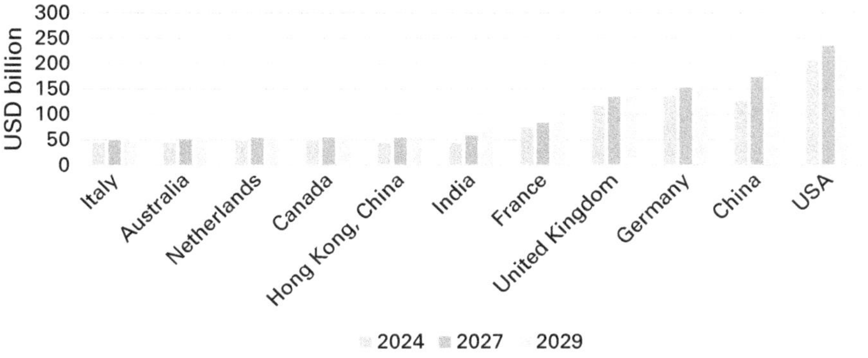

SOURCE Euromonitor International/World Tourism Organization (UN Tourism)
NOTE Value at constant 2024 prices, fixed exchange rates

such as GSTC advocate sustainability measures for different types of travel products and services, a similar approach could be applied to travellers. There will be trade-offs by destinations and travel brands as they grapple with driving the most incremental value to communities, their employees and consumers, while keeping a close eye on carbon emissions. The traditional chasing volume-demand approach has been replaced by chasing high spenders from Asia Pacific to the Middle East. This amounts to short-termism if focusing solely on high spending, even if looking at daily spend per visitor or per trip. A new approach is required for assessing the tourism multiplier effect, shifting away solely from economic benefits of jobs created or revenues generated.

A more efficient way to assess source markets going forward is through a holistic perspective, conscious of all possible negative and positive impacts. Identifying pros and cons can then be discussed with stakeholders. Gap analysis will help to carve out new traveller segments to explore those particularly interested in travelling off-season or to secondary or tertiary destinations. This will need a transition from short-term demand scheduling for 6 to 12 months to long-term demand management. Airlines are especially guilty of near-term planning. The challenge is to move away from 'country pair' thinking; however, this is difficult where geography of origin-destination is integral to the movement of people.

China is forecast to overtake Germany as the second-largest source market spending almost US$200 billion on international travel by 2029, second only to the US on US$255 billion.

TIPS

- To transform to a net zero tourism model, looking beyond traditional source market demographics is key, rebalancing to intra-regional and domestic markets.
- Targeting travellers with more sustainable travel behaviours will become more of a science than marketing without veering into bias and discrimination.
- Aviation lends a higher carbon intensity to long-haul source markets; the trick will be to encourage longer stays.
- Electrification of public transport and the car rental fleet will help mitigate emissions.
- Moving away from a near-term view of airline scheduling would help to maintain consistency of demand.

Identifying traveller segments with passion and positive impact

One sure-fire way of ensuring a robust source of tourism demand and higher spending that is resilient to macro-economic headwinds is by identifying travellers with a passion. Their love of an interest or hobby such as the arts, culture, festivals, wellness, sports, music, celebrities or adventure inspires them, and they are willing to travel to fulfil their interests and aspirations. When it comes to the different generations of traveller segments, it is equally important to understand their motivations, attitudes and preferences.

Adventure travellers boost equity

Adventure travel is highly aligned with sustainable travel, led by the activities of the Adventure Travel Trade Association (ATTA). One area where adventure travellers deliver incremental benefits to local communities and destinations is through the lower levels of leakage. This refers to the amount of tourism revenues remaining in the destination rather than leaving the country. The economic value that adventure travellers bring to a local destination and communities is highly desirable, yet the social and environmental impacts also matter. Three-quarters of an adventure travel trip spending, averaging US$3,000, goes to local suppliers.[17] There is a preference to stay in local accommodation such as eco-lodges, source food locally, use local guides and engage in authentic cultural experiences. Not surprisingly, international institutions, investors, tourism boards and local communities are turning more to adventure travel to transition to a fairer and climate-conscious form of tourism.

Traditionally, adventure travellers come from advanced markets such as North America and Western Europe. Adventure is gaining in popularity in emerging markets such as Asia Pacific. Generation-wise there is a higher proportion of older travellers with 50 per cent aged 41 to 60 years old.[18] The choice of adventure trips is vast, with trekking, hiking, walking, cultural and gastronomy, mountain biking, safaris and wildlife watching being the most popular.

There is an ongoing focus on targeting families, solo travellers, women, 50 plus, LGBTQ+, regional travellers and people with disabilities, showing the diverse appeal of adventure travel. The sector is highly aligned with the SDG agenda where 68 per cent of adventure travel companies have or are pursuing sustainability certification like B Corp or Travelife.[19] This integrated approach – combining the thrill of soft or hard adventure with planet-friendly practices – is what makes adventure travel stand out as a key segment to watch and learn from.

Key consumer trends that are playing out are the desire to go off the beaten path, explore remote destinations and enjoy the joys of slow travel and deeper immersion in a place by staying longer. There is also a strong interest in more customized and personalized trips (see Figure 3.2).

In terms of the adventure travel business community, concerns include safety and security, overtourism and sustainability, while looking to understand the impact of Generative AI. The impact of climate change is already shaping trip itineraries and traveller behaviour, where Europe is increasingly popular with travellers in the shoulder seasons or low season in search of cooler climates, while overtourism is making travellers think twice and finding alternatives to explore instead of saturated locations, so-called 'dupes'. Key motivations for travellers booking adventure travel are to enjoy new experiences, go off the beaten path and travel like a local.

FIGURE 3.2 Top 10 traveller motivations for adventure travel

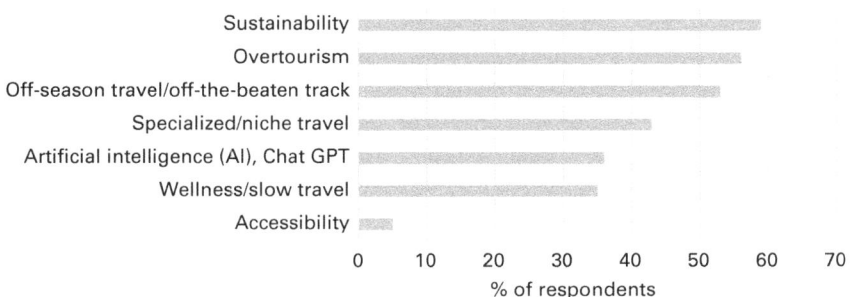

Adventure Travel Businesses: What Trends are Top of Mind in 2024

SOURCE Adventure Travel Trade Association
NOTE Member respondents (n=183) based on consumer demand and bookings in 2023/2024

Savvy trip planning is adopted to ensure that adventure travellers get the most value from their trip, and spend longer in-destination. Safety and security is influencing where travellers feel confident enough to travel. Adventure travellers are onboard with adopting sustainable travel behaviours such as eating and buying from local suppliers (54 per cent), taking public transport (39 per cent) and avoiding overcrowded or environmentally sensitive sites (31 per cent).[20]

Adventure travel intersects with sustainable travel in countless ways where the adventure traveller is high value, low touch with minimal impact. For example, Pura Aventura (UK) is certified B Corp, operating in adventure travel and embraces the no-fly trend offering trips to Spain by train, for hiking, culture and gastronomic experiences.

Adventure also means different things to different people so there is a raft of adventure trips across the seven continents for different fitness levels, from soft to hard adventure, at multiple price points. As seen with the Titan Submersible disaster, operated by Ocean Gate, a select number of high net-worth individuals (HNWI) adventure travellers continue to push the boundaries to explore the furthest corners of the globe.

Expeditions to the Antarctic or Artic poles offer a once in a lifetime opportunity to visit the most remote places in the world for education, wildlife watching and true adventure. Adventure operators like Swoop (another B Corp) offer access to Antarctica that only a few hundred people experience per year, offering activities from camping and kayaking to paddle-boarding. There is a strong educational element to visiting the seventh continent, with visitors encouraged to engage in citizen science and contribute. This type of last-chance tourism, visiting at-risk and fragile destinations, can also bring about negative impacts such as pollution including carbon emissions and noise pollution along with visitors disturbing natural habitats and wildlife such as polar bears.

For nature-based adventures, safari and wildlife watching remains a top passion for adventure travellers. Popular destinations include The Galapagos, already suffering from overtourism and threats to its unique biodiversity with higher tourist taxes introduced and limits imposed. Restrictions are applied to the number of boat and cruise passengers that can visit the island on a daily basis. Backpacking, gap years and flashpacking by older generations continue to attract those looking for longer adventure-inspired trips.

REAL-WORLD EXAMPLE

Evaneos offers a new business model that puts people and places first

Positive impact travel sites like Evaneos help adventure travellers besides other segments to create their tours directly with local travel agents for a more local, sustainable and curated experience. Evaneos, one of the first French tourism businesses to become B Corp, cuts out the middle man dealing directly with local agents and enabling travellers to co-create their trip in line with their 'Better Trip' vision. Through its direct to market model, 85 per cent of the trip value goes to local people, which is an impressive benchmark for others to emulate.[21] EUR3 million has been promised to mitigate the negative impacts of travel and tourism such as offsetting carbon emissions from trips through its carbon offsetting partner, Southpole.

Offering adventure combined with off the beaten track helps to combat overtourism, disperse visitors more fairly and explore undiscovered paths. This equally helps to redistribute tourism revenues for a more equitable share of wealth. Some hiking destinations that are emerging include Mongolia and Madagascar. Evaneos's aim is to encourage all its agencies to be working towards sustainability certification through Travelife to measure their impact and will help 50 per cent directly by funding them.[22] Fifty-eight per cent of Evaneos's agencies are working towards sustainable travel certification which will help consumers to make a better travel choice (see Figure 3.3).

FIGURE 3.3 Evaneos impact findings

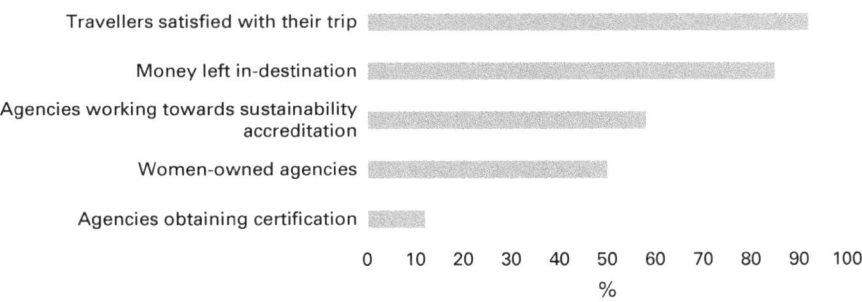

SOURCE Evaneos Impact Report 2023[23]
NOTE 600 agencies, 700,000 travellers

Wellness travellers seek holistic wellbeing

Wellness and wellbeing are megatrends that continue to transform with socio-demographic influences such as the ageing population and the importance of appearance in the age of TikTok and Instagram. Wellness tourism continues to be popular as people increasingly look to travel brands and destinations to improve their holistic wellbeing while away from home. By 2027, the Global Wellness Institute forecasts that the market for wellness tourism will reach US$1.4 billion with a 16.6 per cent CAGR for 2022–2027.[24]

The wellness travel market is highly diverse, disrupted by a myriad of trends from nutrition, physical, emotional and spiritual, at different life-stages to emotional states such as dealing with the loss of a loved one. Technology, cutting-edge diagnostical tools such as AI and sustainability are all having an impact. There are different approaches in what wellness travellers are after, either hardcare or softcare, where the former refers to high-tech and medical versus a simpler and emotional solution.

With the number of people aged over 65 doubling to 1.6 billion by 2050, wellness travellers will also age and increasingly look for wellness treatments combined with travel experiences.[25] These older wellness seekers will also have more time and funds after leaving the workplace. With a growing emphasis on ageing well, longevity techniques and treatments have made a big impact on wellness in the past few years, amounting to US$610 billion by 2025.[26] Holistic wellness encompasses mind, body and soul to boost longevity. The most famous anti-ageing and health resort in the world is Clinique La Prairie in Switzerland, combining genetics, medicine and innovative technology. Others, like Six Senses in Portugal and Switzerland, offer personalized holistic anti-ageing programmes, combining neuroscience, healthy eating, ancient knowledge and fitness focusing on areas such as sleep and providing biohacking for long-term regeneration.

Wellness travellers looking to age well will naturally be interested in regenerative, nature-based and holistic wellbeing travel businesses that deliver personalized diets and healthy food and drinks. Treatments such as cold-water bathing, ice baths or cryotherapy are gaining in popularity.

Young people are increasingly experiencing loneliness due to social media that has been proven to cause feelings of solitude and depression, which opens up wellness tourism to those seeking to reconnect with others, although there are equally silent retreats for solo travellers who seek solace away from the daily noise where there is greater autonomy and self-determination.

Millennials are highly experiential and prioritize travel, but they are ageing and hitting 40 years old. With this new-found self-awareness of their own mortality, they are looking for self-care, with 61 per cent travelling to focus on personal wellness and 60 per cent looking for a hotel with a spa.[27] Where travel businesses align with consumers' expectations in terms of quality experience and shared values, the rewards will be plentiful where 79 per cent of Millennial travellers care more about the right trip than the cost of the trip.[28]

Luxury travellers reimagined

There will always be a market for luxury travel, with money enabling exclusive access to behind the scenes, being treated like a VIP. The images of excess, overindulgence and waste that stigmatize luxury will slowly be replaced with a more human and emotional experience. Original tenets of luxury travel continue to remain such as curation, personalization, authenticity and the search for the unique and new.

REAL-WORLD EXAMPLE
Black Tomato rejects the bucket-list for unfiltered travel

Luxury travel brands like Black Tomato, a member of Virtuoso, exemplify how luxury travel operators aim to inspire their travellers, offering curated itineraries from the Amalfi coast, Iceland, the Okavango desert, hiking in Patagonia to wellness retreats in Asia. The goal is for immersive and unfiltered travel experiences, aimed at evoking a feeling of going deeper into a place and its culture. Their motto is to be curious, thoughtful and humble, and encourage their customers to follow this approach.

For the more adventurous looking for a challenge, Black Tomato offers a 'Get Lost' option where the traveller does not know where they are going or even what they need when they arrive. The company caters for families, solo travellers, couples and groups, applying a regenerative approach to the communities and destinations it operates in. Brands like Black Tomato reject the bucket-list mantra.

TIPS

- Resisting the temptation of the bucket-list is an important step in sustainable travel.

- For every TikTok or Instagram post of exciting travel experiences and attractive destinations, increasingly there will be counter posts to show where these are causing negative impacts.

- Travel businesses can shine a light on unsustainable practices by using authentic imagery and being open about downsides rather than showing only glossy images of destinations.

- Redefining travel in a sustainable and human-centric way is long overdue.

Social media and web 3.0 have democratized luxury, opening it up to younger people who seek to have a more digital, direct and participatory engagement with the brands they love. Luxury fashion brands like Gucci, Burberry and Louis Vuitton have moved into the metaverse – the next generation of the internet. This opens up opportunities for old-school luxury to reinvent itself and also creates openings for a new school of luxury that is more inclusive and diverse, with lower barriers to entry. Breaking down traditional barriers of age, income, sex, race, pivots the luxury sector to a more open future.

The way luxury travellers engage with destinations and their communities is transforming to a more conscious, purpose-driven engagement. Climate change and the net zero transition are everybody's concern, where money will not protect the wealthy from its impacts.

Luxury travel in its very nature is more carbon intensive, using more resources per traveller. Just 1 per cent of the population causes 50 per cent of carbon emissions from aviation, and private jets are five to 14 times more polluting.[29] Environmental charities like Greenpeace are calling for private jets to be banned to tackle climate injustice and inequality.[30]

For HNWI, their attitudes and behaviours are aligned with the need for sustainability where 75 per cent agreed that they try to tread lightly and reduce their carbon footprint.[31] The figure was even higher for HNWI Millennial men (80 per cent) and women (79 per cent), illustrating how younger generations are onboard.[32]

Big and small luxury operators are shifting their focus; some are moving faster than others to the circular economy with regenerative benefits.

Reaching net zero emissions is a top priority with carbon-neutral named as the new gold standard for luxury hotels, cruise and tour operators.[33] Luxury travellers are looking to have a positive impact, looking for immersive cultural and nature-based experiences and engaging in truly local and co-created travel products with communities.

REAL-WORLD EXAMPLE
Kalukanda House – redefining luxury for an inclusive and emotional connection with place and people

Dee Gibson calls herself an accidental hotelier who came to hospitality through chance. With her background in 'emotional design', she has established Kalukanda House in Sri Lanka. Her goal with the luxury villa is to create an emotional connection between the visitor, community and the country through design that uses local antiques and materials, offering local-led experiences.

A circular economy approach is integral to the business, ensuring that local communities receive the maximum revenue from the trickledown effect, where there are high levels of hardship and food insecurity in the country. Environmental metrics are important but it's more than that, it's about cultural exchange, social mobility and positive social impact.

Dee is working to redefine luxury which can be seen as a 'dirty word and elitist'. She is keen for luxury to be redefined to be more inclusive, regenerative and not defined by budget. Equally important is to participate in slow luxury travel, spending longer in-destination and visiting a couple of places, which compensates for the long-haul flight. Empowering travellers to be more conscious of where to go and how they spend their money is key to a sustainable future, mitigating carbon and contributing to the circular economy.

Dee believes that destinations like Sri Lanka that have a history built off trade routes on the Silk Road will continue to invest in tourism development, and not reinvent the wheel. Businesses like Kalukanda House are helping the necessary transformation at grassroots level.

Kalukanda House hosted a three-day event for 100 per cent Sri Lankan female Creative Founders, the first of its kind in the country for guests and the community. The Hera PROJECT X event included workshops and brought together women-owned businesses from arts, crafts such as lacemaking, fashion, jewellery design, travel and photography. One powerful story is about young women from a Colombo shelter supported by Kalukanda that fled Myanmar under life-threatening circumstances and were inspired by the creative design workshop. More young women now attend jewellery workshops in the capital with the same designer to learn the necessary skills so that they can live independently and earn a living, and 100 per cent of the workshop funds will go back to the artisans while providing guests with an unforgettable experience. This is female empowerment in action.

Technology also plays an important role in the circular economy in Sri Lanka where blockchain is being introduced by one national textile business to provide transparency on how much artisan handloom weavers and seamstresses are paid. The importance of sharing stories, storytelling and education is wrapped up into the business ethos, changing the narrative about Sri Lanka and its people.

'For us, luxury travel is about giving our guests an emotional feeling and an amazing experience. It is not about looking at the country through the glass screens, but interacting and having a two-way conversation for cultural exchange.

In the future, we need more people travelling as what would happen from a human perspective if they didn't?'

Dee Gibson, CEO, Kalukanda House

TIPS

- Travel is about emotion and creating a feeling about a place and its people.
- When taking a holistic and circular approach to running a travel business, the ripple effects are widespread, lifting up everyone in the chain.
- Changing the narrative: social media and the media in general need to transform to align with the sustainable transformation in travel.
- Media companies like Netflix should work in partnership with destinations to ensure the sustainable management of visitor flows after hit shows are released such as *Emily in Paris*.

Music, festivals and cultural events

Under cultural tourism are popular activities such as attending concerts, festivals and music events. Mega concerts by global artists such as Beyoncé, Taylor Swift, Coldplay, K-Pop groups or U2's residency at the Sphere, Las Vegas, continue to attract visitors and fans. Equally popular are music festivals from Coachella (US), Lollapalooza (US), Glastonbury (UK) and South by Southwest, mixing film, music and tech (US).

Cultural festivals such as the Edinburgh Festivals have been running for eight decades as proven drivers of economic growth. The 11 festivals attracted 700,000 attendees with an economic impact of GBP367 million to the wider economy in Scotland.[34] The length of stay has extended and visitors are increasingly travelling beyond Edinburgh to other parts of the country to help drive the value of festival tourism. However, when considering the cost-benefits of visitors' negative environmental impacts from their journeys to the festivals, the value per type of visitor (long haul, resident versus domestic) would look very different.

In cities like London, cultural tourism is a big draw where the city used the 2012 Olympics and Paralympic Games mega event to rebrand its destination image. Live events were the third popular reason for visiting, after the cultural experience and visiting famous landmarks and attractions, across all age groups.

Cultural tourism has struggled with maintaining its integrity in the face of ever-increasing homogenization of culture and media through technology. Tourism also plays a role as international visitors bring with them their own preferences and expectations which can influence and diminish local experiences, food and shopping. This leads to the same-ification of experiences across countries, sharing a lack of diversity and authenticity. To counteract the trend for same-ification and commoditization, DMOs are co-creating cultural experiences with local residents and communities to ensure that cultural integrity is preserved.

However, as Scott Wayne, SW Associates, LLC – Sustainable Destinations mentioned, there is a lack of good data on understanding the cultural tourism sector which has great potential to drive equity and inclusion, especially for women.

REAL-WORLD EXAMPLE

Swiftonomics – influence of pop culture on travel

Taylor Swift's Eras tour (2023–2024) provoked a big uplift in tourism activity in cities across the world driven by her fans' desire to see their favourite singer/celebrity live.

This economic boost has been defined as 'Swiftonomics', with increased spending on hotels, transport, food, restaurants and shopping. In the US alone, the direct economic impact is valued at US$5 billion, with a multiplier effect of more than US$10 billion in terms of total economic impact.[35] The economic impact argument is clear.

However, from a sustainability perspective, Taylor Swift's use of private jets in general and during her Eras Tour has garnered much media attention about not just celebrities' carbon footprints but also the climate impact of private jets. Swift's response was to purchase double the amount of carbon offsets to appease her critics. However, the process of carbon offsetting is regarded as greenwashing by many due to the inefficiencies of such schemes. Role model celebrities such as Taylor Swift influencing younger generations should also be encouraged to engage and contribute to shaping climate-positive behaviours.

TIPS

- Taking a holistic view of visitors' impacts in terms of environmental and social would give a more rounded view of the cost-benefits of cultural tourism.

- There is a need for an ongoing cultural renaissance that DMOs can champion.

- Tapping into hyper-local, co-created experiences between locals and DMOs and DMCs ensures authenticity and provenance.

- Digital tools help to drive discovery of places and communities off the beaten path such as audio-tours.

Sports fandom's love of travel

Sports tourism is defined as the engagement of visitors participating in sports activities or spectating sports, with the segment worth 10 per cent of global tourism, forecast to grow by 17.5 per cent over 2023–2030.[36] Sports fans are renowned for their passion and enthusiasm for their teams and love to travel to immerse themselves in the sports experience.

In 2024, there is much higher engagement from younger cohorts, with 67 per cent of Millennial and Gen-Z respondents, versus 58 per cent of all respondents, that are interested in travelling for sporting events in 2024.[37] It's not just the young who love to travel for sport. In Scotland, the Scottish Masters Hockey team including the super veterans with their wives in tow, played games in destinations worldwide including Greece, Hong Kong and Australia over the years.

Sports mega events can be a double-edged sword for countries, hosting the FIFA World Cup or the Olympics and Paralympics Games. Every destination wants to be the next Barcelona, yet it can often be a highly costly investment with no long-term benefits as seen with South Africa who played host to the 2010 FIFA World Cup. The legacy entailed several empty stadia and debt. The year of a mega event can also cause displacement from traditional source markets, that may avoid the key tournament weeks to avoid high prices and the crowds.

For Real Madrid, the Bernabéu Stadium is more than a stadium. It is a visitor attraction plus a leisure, sports and cultural entertainment space with a recent investment of EUR360 million in its transformation.[38]

EXAMPLE
2024 Paris Olympics and Paralympic Games goes carbon-conscious

For the 2024 Paris Olympics and Paralympic Games, the aim was to cut carbon emissions by half compared to previous games and serve more plant-based meals to lower CO_2 through its Food Vision leveraging the country's famous gastronomy and creative innovation. The food programme involved providing twice as much plant-based food, 100 per cent certified, 80 per cent French produce, zero waste among other initiatives.

Mainstream sustainable makeover

The leisure mass market continues to come under heavy fire as anti-tourism protests in key destinations like the Canary Islands unfold. Leisure tourism that has ballooned out of control in island resorts and city break destinations like Barcelona and Kyoto has led to residents and community groups uniting to say enough is enough. For residents, the balance has tipped to where the negative impacts of tourism outweigh the benefits.

The mass tourism model is broken and there is a limit, as the residents of the Canaries are calling for an end to this exploitation and the introduction of an eco-tax. There is anger at large-scale hotel developments, short-term lets and foreigners buying property forcing locals out due to being unable to afford rents. For many leisure travellers they will be shocked and surprised to hear that the holiday that they are looking forward to is causing so much upset and frustration to the communities they visit.

The package holiday market in Europe has mushroomed in the past several decades typified by cheap flights, deals, discounts and all-inclusive offers so that costs are fixed up-front for spending on food and drink. Northern Europe source markets like the UK, Scandinavia and Germany are the most avid package holiday travellers, heading to the Mediterranean for sunshine, rest and relaxation. Traditional high-street travel agents and tour operators have been disrupted by the online giants like Airbnb, Booking.com, Expedia and Trip.com as well as low-cost airlines (LCCs) like easyJet and Ryanair. Due to the low margins and digitalization, many players have not survived such as Thomas Cook going bankrupt in 2019 after 178 years in business.

REAL-WORLD EXAMPLE

TUI transforming the mass market for a sustainable future

TUI has navigated successfully the digitalization of the leisure travel market while embracing sustainability. It served 19 million customers in 2023, hedged by its ownership of assets with 360 hotels, 16 cruise ships, over 45,000 experiences and 126 aircraft as part of its integrated business structure.[39] To continue to unlock value, it is launching new products and targeting new customer groups underpinned by technology and the drive for personalization. New products include dynamic packaging, while unbundling to sell ancillaries, car rental, single components such as flights, hotels and tours. In terms of new segments, it has its sights on 'smart tanners' (looking for sun & beach), 'senior service' (baby boomers and silent generation), 'home and aways' (domestic/international), 'travelistas' (looking for travel experiences with cachet) and 'energized adventurers' (into sports and adventure).[40]

TUI has been on the sustainability pathway for over a decade. It has embraced STBi targets aligned with the SDGs, aiming to be net zero and a circular business by 2050. As part of its emissions roadmap to 2030 are ambitious targets for emissions reductions of 24 per cent for airlines, 27.5 per cent for cruise and 46.2 per cent for hotels and resorts.[41] TUI's positive financial results demonstrate that value creation and sustainability combined are a resilient business model for the mass market.

There is still a last-minute deals market for packages and flights, fuelled by budget players targeting the most price-conscious, and metasearch platforms like Kayak and Google. However, with destinations turning their back on overtourism and consumers more eco-conscious, the long-term outlook for the budget sector is uncertain. As travel transitions to a quality tourism model, there will be a hollowing out of the bottom of the market. Travel businesses need to work even harder to be transparent about their positive and negative impacts so that their consumers can make the right choice.

Offering flexible payments such as pay in instalments or buy now, pay later (BNPL) enables consumers from lower to middle income households to continue to afford their dream holidays. For example, BNPL player, Klarna, partners with travel brands such as Expedia and Trip.com.

The all-inclusive package is deemed the most unsustainable, with count-less research on the negatives inflicted on the local economy, environment and community. Key weaknesses include the low spending in-destination on services and goods as visitors remain onsite at resorts (so-called enclave tourism), the significant carbon footprint from energy and waste, and lack of interaction between visitors and the host communities. It is a contentious issue with debate about whether such business models should be banned as was attempted in Gambia. Even sustainability-led leaders like TUI have their First Choice brand, which is 100 per cent all-inclusive, showing that sustainable principles can be applied to the all-inclusive resort model. The key is to ensure that there is local sourcing, community-led experiences and a reduction in the carbon footprint. Opening up all-inclusive resorts and integrating the local community into the experiences on offer through cultural exchange will deliver greater understanding and connection, inspir-ing repeat visits.

Marriott Bonvoy launched all-inclusive luxury resorts post-pandemic in the Caribbean, Costa Rica and Mexico, illustrating that there is demand for all-inclusive from its loyalty members. However, the question is whether it will slow its progress to achieve its near- and long-term carbon targets?

TIPS

- Develop curated and personalized travel services and experiences as travel becomes increasingly experiential and interests-led, where consumers look for brands with shared values.

- For the leisure traveller, it is not just about price – it's about value for money and the best experience.

- People will pay more where they can enjoy their passions and interests, and fulfil their aspirations with sustainability built-in.
- As the leisure market increasingly premiumizes, travel businesses must enable lower- to middle-income families continued access to travel experiences through flexible payments and other inclusion measures.
- Embracing locally sourced food, drinks, artisanal products and crafts delivers authenticity for visitors and empowers local communities for sustainable transformation

Cruise controversy over its carbon footprint and impact on destinations

Cruise is one of the most controversial talking points in travel where the sector accounts for only 1 per cent of the travel industry's global carbon emissions, amounting to 27 million tCO2e emissions pre-pandemic.[42] However, it is more carbon intensive per passenger than even aviation, at 0.91 tCO2e emissions per cruise passenger compared to 0.22 tCO2e per passenger for air passenger pre-pandemic.[43] Cruise clearly has a disproportionate effect on the environment and the oceans, is highly polluting, contributes to overtourism, and tends to not drive equitable tourism spending. Compared to land-based visitors, cruise goers produce eight times more carbon emissions.[44]

The cruise sector has exceeded its peak pre-pandemic levels, carrying almost 32 million passengers in 2023 amounting to US$138 billion, and potentially exceeding 40 million by 2027 due to increased capacity.[45] Each passenger spent on average US$750 in local destinations over the course of a seven-day cruise.[46]

The future cruiser is pivoting to the Millennial and Gen Z away from the traditional Baby Boomers, Silent Generation and Generation X. Source markets for cruise travellers have an average age of 46, and skew towards advanced regions – US, Germany, UK – however, cruise is gaining popularity in Asia Pacific especially China, Australasia and Latin America.[47] The Caribbean and Mediterranean are the most popular cruise destinations and have spent decades investing in ports and infrastructure. While mass market cruises offer value for money for consumers, there is strong interest equally in luxury and ultra-luxury cruises that tend to use smaller vessels.

Key consumer trends are solo travellers, multi-generational, accessible travellers as well as appealing to younger demographics. Large cruise companies like Royal Caribbean and Carnival are keen to ensure that they offer inclusive and accessible cruise experiences.

With the ever-constant search for new experiences, the major cruise lines – including Royal Caribbean, Carnival, Virgin, Holland America, Norwegian and Disney – have invested in private islands, building exclusive resorts in the Caribbean. For destinations, this is a challenge, especially where local residents have relied on cruise ships for trade. On the flip side, this removes the challenges of overtourism. While cruise goers benefit from the all-inclusive services at the private island beach resorts and clubs, with food, drink, pools and entertainment, in ports like Nassau and Freeport (Bahamas), it starves the local communities of much-needed revenue.

In ports such as Key West, Florida (USA), residents have called for restrictions on cruise ships limiting the port to one ship per day with a maximum of 1,500 passengers. However, there has been much controversy in Key West where the ruling was overturned by Florida's Governor DeSantis. Concerns about damage to the coral reef persist.

The future of cruise lies in its transition to net zero by 2050 as per international shipping targets led by Cruise Lines International Association (CLIA). Yet the future is uncertain where concerns such as pollution from toxic waste and sewage, noise and air pollution, damage to biodiversity, marine life and the oceans continue to surround the sector. With the future cruise goer being younger and more planet-conscious, the cruise sector will need to sharpen up its act to continue to appeal to the future traveller.

TIPS

- Cruise is highly controversial and requires a holistic approach including cost-benefits.

- It is critical to take into account the voice of local communities.

- Rather than buying private islands for resort development, investing in existing infrastructure would help to drive local benefits and reduce revenue leakage.

- Travel businesses can help influence change by working with smaller cruise companies following sustainability pathways.

Beyond niche to drive inclusion

There are countless different ways to segment travellers, but, in the future, it will be a case of providing as much personalization as possible. Firstly, this means ensuring that travel is inclusive and welcoming to all, regardless of age, race, sex, income, ability, faith or nationality.

The accessible travel segment continues to attract more attention as diversity, equity, inclusion and accessibility (DEIA) becomes more integral to businesses' ESG reporting and the need to become more people-centric. The accessible traveller also delivers higher spending with a high multiplier effect where for every accessible traveller, there are at least up to two more people in the group if not more.[48] Hence the purple pound offers much potential.

For many travel businesses, unfortunately, accessibility is deemed a matter of compliance. Yet, the returns on investment are high when delivering an open-to-all experience and adopting a universal design approach to the built environment. Training is a critical component of delivering a seamless and accessible service experience.

Like with all segments, there are important differences when targeting accessible travellers, where disabilities can be physical, sensory and/or invisible impairments such as cognitive or mental health. At least 15 per cent of the world, amounting to 1.2 billion people, live with a disability and face discrimination and barriers to living their best life every day.[49]

REAL-WORLD EXAMPLE
Sage Inclusion elevating inclusion for everyone in travel

Sage Inclusion is a global leader in accessible travel, established by John Sage, CEO, to ensure that accessibility and inclusion is central to travel businesses. Sage Inclusion's work includes travel services for accessible travellers, professional training, strategy and certification, partnering with CLIA, WTTC, Global Business Travel Association (GBTA), SHA and ACI. Certification covers five different areas: mobility, visual, hearing, cognitive and allergy. The range of travel businesses covered includes hotels, cruise, airports, airlines, transport and tours. The Sage Certification has also adopted three international ISO standards including the Americans with Disabilities Act (ADA) which is the minimum that businesses should aim for. Its hospitality assessment tools for hospitality are free, so there are no barriers to adopting more inclusive standards.

The benefits of adopting inclusion and accessibility go beyond compliance to meet regulations. It drives customer loyalty, revenue opportunities and has positive social and equitable impact. Fundamental to Sage Inclusion's work are accessible facilities including the seven principles of universal design, accessibility documentation and accessible customer service. The ISO for accessible travel for all is seen as a key foundation where there is no need for adaptation, and the product/service is open to all from the outset. This ensures that accessible travel does not operate in silo but rather is an integral part of the travel industry.

Equally, it is important to see people with disabilities as people first and foremost, who have specific needs. Available and up-to-date information is essential for People with Disabilities (PwD) to travel, hence the importance of the planning stage of the trip. Measurements and photographs play an important role. It is also vital that every stage of the journey is accessible, with no breaks in the chain. Without the required information, accessible travellers will not travel.

Starting with getting the basics right such as information provision is worth the effort, especially to cater better to a high spending and loyal traveller market and ensure travel is truly inclusive.

'Accessible travellers are a significant and growing demographic that the travel industry cannot afford to overlook. Taking an inclusive approach to experiences and services not only ensures compliance with accessibility regulations, but also opens up new market opportunities and enhances overall customer satisfaction. By prioritizing accessibility, businesses can create memorable experiences for all travellers, fostering a more inclusive and welcoming environment for everyone.'

John Sage, CEO, Sage Inclusion

Inclusion is a broad and welcoming church

The LGBTQ+ community and the power of the pink pound are another important source of high tourism spending. The International Gay and Lesbian Travel Association (IGLTA) provides information on inclusive destinations and travel brands, along with safety information. The IGLTA has rolled out a new accreditation scheme for hotels to reassure guests that the service is welcoming and safe based on select criteria. Different destinations and countries around the world have different attitudes and laws, so it is vital that LGBTQ+ travellers have the most relevant information so they can make an informed choice on where to travel. Transgender people are the most at risk of discrimination and intolerance. IGLTA partners with Destination Pride which is an online platform that ranks the LGBTQ+ friendliness of countries based on metrics such as laws, rights and social sentiment.

The membership of IGLTA is broad, including luxury travel, adventure, tour operators such as The Travel Corp, Disney, Belmond among hundreds more as well as highlighting LGBTQ+-owned businesses. MisterB&B is another source of inclusive travel information from its community of 1 million people. It offers hidden gems, tips, recommendations, meet-ups with locals as well as providing short-term stays with hosts Airbnb-style, travel companions or LGBTQ+-friendly hotels.

Black travellers are an important segment that are often not catered for. Following the rise of #BlackLivesMatter and the murder of George Floyd, there is a strong interest in cultural and historical experiences that showcase Black history and culture, past and present, in an authentic way. NGOs like Black Travel Alliance (BTA) are working with the travel industry and content creators to change the narrative, focusing on alliance, amplification and accountability. Their Generosity Movement promotes the importance of storytelling and representation. Brands working with BTA social media creators like bloggers and influencers include Hurtigruten showcasing the Galapagos for adventure travel experiences, Portland and Barbados.

Travel operators and communities such as BlackGirlsTravelToo (BGTT) offer tailored experiences to the Caribbean, Asia, Europe and Africa, for cultural immersion and voluntourism. Such communities have been created out of a gap in serving women of colour and enabling them to enjoy the trip of a lifetime. The ethos of BGTT is empowerment, journey of self-discovery, inclusion and sustainability.

Halal tourism is an important emerging sector of travel, expected to generate US$225 billion and 230 million arrivals by 2028.[50] The global Muslim population is growing fast and expected to reach 2.3 billion by 2030 with an average age of 25.[51] Providing halal-friendly services in-destination, at the hotel and airport are vital including halal food and dining, as well as access to prayer facilities for an inclusive travel experience. According to the Global Muslim Travel Index, Singapore ranked first for Muslim women-friendly destinations, followed by Taiwan and Japan for non-Organization of Islamic Cooperation (OIC) countries while Malaysia, Indonesia and Qatar topped the OIC destinations.[52]

TIPS

- Embrace diversity and inclusion principles for products, services and experiences.
- Inclusion is about belonging, feeling welcome and being treated with respect.
- There is huge opportunity for co-creating innovative products and services that promote diversity and inclusion.
- Authentic storytelling by real people helps to boost trust about inclusion and safety.

Business traveller transformation

Business travel is a segment that continues to undergo massive disruption and transformation. Its very existence was called into question at the height of the pandemic when businesses and employees pivoted to remote working and virtual meetings on Zoom. However, recovery is underway. According to the GBTA, business travel spending is forecast to hit US$1.8 trillion by 2027, up from US$1.4 trillion in 2024 returning to peak levels where an average business traveller spends US$1,018 per trip.[53]

Business travel is more susceptible to macro-economic and geopolitical threats than leisure travel. It is increasingly under the scrutiny of businesses' ESG targets from investors, employees and consumers. Business travel has a higher carbon footprint if booking business-class and first-class flights (more than three and four times higher $kgCO_2e$ respectively than economy for a UK long-haul flight).[54] Corporate travel preferences tend towards more upper scale and luxury hotels with higher carbon intensity. However, only 19 per cent of GBTA respondents state that climate impact and sustainability were top concerns in 2024, higher in Europe at 41 per cent.[55]

With the new EU Corporate Sustainability Reporting Directive (CSRD) that requires large and listed companies to report regularly on their impacts on the environment and society from 2025, there will be even sharper scrutiny of business travel emissions. The CSRD includes the new concept of double materiality, considering inside and outside impacts, in terms of financial, environment/people related, or both. Large companies have identified that the business travel of their employees and leadership teams is a major contributor to their carbon emissions, and targeted for reducing non-essential travel. Climate conscious employees are also taking a moral stand by choosing no-fly alternatives to business travel.

Blended travel (also known as bleisure, mixing business and leisure) was accelerated post-pandemic as businesses refreshed their policies to allow employees to work from anywhere, boosting workcations and digital nomads, with 62 per cent of business travellers blending work and leisure more often.[56] Countries have actively targeted the digital nomad through digital nomad visas, with countries like Spain and Argentina being popular.

The blended traveller is a paradox as they are spending longer in-destination and travelling off-season, however they are also named as a contributor to overtourism in destinations like the Canary Islands and Costa del Sol, staying in short-term lets and pricing residents out of the housing market.

Summary

With the rise of emerging market travellers, in line with increased disposable income and connectivity, the future face of travel demand will be ever more diverse. Regions like Asia Pacific, the Middle East and Africa are poised for strong growth, an untapped source of new travellers. Yet travellers will not be defined by their country of residence but overwhelmingly by their passions, values, beliefs and lifestyles.

Travel businesses will need to ensure that they take an inclusive approach to their products and services so as to ensure that everyone is welcome. With the pressures to transform to a net-positive travel and tourism model, greater scrutiny will be applied to achieving the optimum mix of demand, bearing in mind positive and negative impacts.

For a travel business to be truly sustainable, it is vital to tackle the unsustainable segments of all-inclusive holidays and resorts, city breaks and cruise that contribute to mass tourism effects. Business travel and first-class travel will also need to be addressed due to their significant carbon footprint. It will be necessary to take a bold stance and correct unsustainable practices that do not deliver holistic benefits. This speaks to the need to work with partners across the travel supply chain to drive resilience, act with purpose and pursue a pathway to sustainable transformation.

The big question is whether destinations and local communities will want to receive the future traveller? The risks may soon outweigh the cost-benefits as already seen in destinations from Bhutan to Amsterdam.

If steps are not taken to address the needs of local communities, the desire to travel will be unmet. Anti-overtourism levers like taxation will further drive up prices, forcing many lower- to middle-income consumers to forgo their travel aspirations due to prohibitive costs. For secondary and tertiary destinations that are less well known and less well trodden, this will be an opportunity to tap into.

Faced with quotas, bans and high taxation, the freedom to travel would be massively curtailed, threating the livelihoods of hundreds of millions of people who depend on tourism for employment and a fairer life. Travel would end up being the preserve and privilege of the wealthy with the means to pay, perpetuating inequality.

Finding innovative digital and sustainable solutions will be fundamental to ensuring travel remains inclusive, and open to all especially as an enabler of peace and understanding about different cultures and beliefs. Now, more than ever, greater empathy is required in an ever-polarized world.

Endnotes

1 Our World in Data (2022) from UN, ourworldindata.org/grapher/population-with-un-projections (archived at https://perma.cc/99EQ-UNB8)

2 Our World in Data (2024) from UN, ourworldindata.org/grapher/population-by-income-level (archived at https://perma.cc/8QLQ-76GK)

3 European Comission (2020) knowledge4policy. ec.europa.eu/ (archived at https://perma.cc/7GXB-TGLD)

4 Institute for Economics and Peace (2023) www.economicsandpeace.org/wp-content/uploads/2023/12/ETR-2023-web.pdf (archived at https://perma.cc/J8G6-9WZ4)

5 Institute for Economics and Peace (2023) www.economicsandpeace.org/wp-content/uploads/2023/12/ETR-2023-web.pdf (archived at https://perma.cc/DJ2D-XC8X)

6 Euromonitor International/UN Tourism (2024)

7 Financial Express (2024) www.financialexpress.com/policy/economy-india-to-have-billion-plus-middle-class-by-2047-study-3157931/ (archived at https://perma.cc/3A59-B2GE)

8 Ujjain (2024) ujjain.nic.in/en/culture-heritage/ (archived at https://perma.cc/S776-C4ZX)

9 MakeMyTrip (2024) promos.makemytrip.com/MakeMyTrip%20Orchestra-A4-new-220424.pdf?open=outside (archived at https://perma.cc/QAF3-E9DB)

10 Ibid

11 Aerospace Technology Institute, Fly Zero (2022) www.ati.org.uk/wp-content/uploads/2022/03/FZO-CST-REP-0043-Market-Forecasts-and-Strategy.pdf (archived at https://perma.cc/9SMF-U683)

12 World Economic Forum from Brookings Institution (2021) www.weforum.org/agenda/2020/07/the-rise-of-the-asian-middle-class/ (archived at https://perma.cc/UD6A-A6EX)

13 UNHRC (2024) www.unhcr.org/refugee-statistics/ (archived at https://perma.cc/2Y5P-MHH6)

14 Institute for Economics and Peace (2020) www.economicsandpeace.org/wp-content/uploads/2020/09/Ecological-Threat-Register-Press-Release-27.08-FINAL.pdf (archived at https://perma.cc/5TZV-72FG)

15 Deloitte (2023) www2.deloitte.com/sg/en/pages/about-deloitte/articles/new-deloitte-insight-eco-consciousness-in-asia-pacific-on-the-rise-urging-accelerated-adoption-of-sustainable-products-and-system-level-solutions.html (archived at https://perma.cc/F5EG-P7ZR)

16 Euromonitor International/UN Tourism (2024)

17 ATTA (2023) cdn-research.adventuretravel.biz/research/64b9e42a905226.17033554/ATTA-Snapshot-Trends-2023-Report.pdf (archived at https://perma.cc/34MW-NW9D)

18 Ibid

19 Ibid

20 World Nomads (2023) www.worldnomads.com/in-the-news/media-releases/2024-travel-trends (archived at https://perma.cc/DBX9-ZFNM)

21 Evaneos (2024) bettertrips.evaneos.com/fr/ (archived at https://perma.cc/TX83-3XL5)

22 Ibid

23 Evaneos (2023) issuu.com/evaneos/docs/evaneos-impact-report-12_en_us_2023.09_20_1_ (archived at https://perma.cc/RG6V-PLSR)

24 Global Wellness Institute (2023) globalwellnessinstitute.org/press-room/press-releases/globalwellnesseconomymonitor2023/ (archived at https://perma.cc/C5MU-USFM)

25 UN (2023) www.un.org/development/desa/dspd/wp-content/uploads/sites/22/2023/01/2023wsr-chapter1-.pdf (archived at https://perma.cc/5KCJ-68M9)

26 European Spa Magazine (2024) europeanspamagazine.com/global-wellness-summit-predicts-10-wellness-trends-for-2024/ (archived at https://perma.cc/3H2N-NBB4)

27 CNBC (2023) from American Express, www.cnbc.com/2023/03/27/millennials-are-turning-40-and-theyre-changing-travel-as-we-know-it.html (archived at https://perma.cc/K4KA-NDK9)

28 American Express (2024) Global Travel Trends Report

29 Stefan Gössling, Andreas Humpe (2020) www.sciencedirect.com/science/article/pii/S0959378020307779?via%3Dihub (archived at https://perma.cc/PTY8-NQMJ)

30 Greenpeace (2024) www.greenpeace.org/international/act/ban-private-jets (archived at https://perma.cc/7HUX-7F26)

31 Knight Frank (2024) www.knightfrank.com/wealthreport/2024-03-05-the-age-of-change (archived at https://perma.cc/C9Z9-E7Z3)

32 Ibid

33 Virtuosos (2023) www.virtuoso.com/travel/articles/trend-report-what-we-learned-at-virtuoso-travel-week-2023 (archived at https://perma.cc/YR39-HJW5)

34 Edinburgh Festivals (2023) www.edinburghfestivalcity.com/assets/old/Edinburgh_Festivals_Impact_Study__digital__original.pdf?1687855168 (archived at https://perma.cc/FKA6-7CQ7)

35 US Travel (2023) www.ustravel.org/news/taylor-swift-impact-5-months-and-5-billion (archived at https://perma.cc/J4EA-ZDQY)

36 UN Tourism (2023) www.unwto.org/sport-tourism#:~:text=What%20it%20is%3A%20Sports%20tourism,activities%20of%20a%20competitive%20nature (archived at https://perma.cc/Q9J7-P4MM)

37 American Express (2024) www.americanexpress.com/en-gb/travel/discover/get-inspired/global-travel-trends-en-gb (archived at https://perma.cc/X6UM-MBBT)

38 Sixth Street (2022) sixthstreet.com/investment_announce/strategic-partnership-between-real-madrid-sixth-street-and-legends/ (archived at https://perma.cc/EC5Q-XPEJ)

39 TUI Group (2024) www.tuigroup.com/damfiles/default/tuigroup-15/en/investors/6_Reports-and-presentations/Presentations/2024/20240312_TUI-Investor-Presentation_Barclays-Leisure-Transport-Conf_vf.pdf-ec968bccaf13735c361ac99ca6ca84c5.pdf (archived at https://perma.cc/L4BR-DC8N)

40 Ibid

41 Ibid

42 WTTC (2021) wttc.org/Portals/0/Documents/Reports/2021/WTTC_Net_Zero_Roadmap.pdf (archived at https://perma.cc/3TBN-HEL5)

43 CLIA (2024) State of the Cruise Industry Report 2024, IATA (2023) Global Outlook for Air Transport, WTTC Net Zero Roadmap (2021)

44 Friends of the Earth (2023) foe.org/news/cruise-passengers-carbon/ (archived at https://perma.cc/ZH72-T3Y5)

45 CLIA (2024) cruising.org/-/media/clia-media/research/2024/2024-state-of-the-cruise-industry-report_041424_web.ashx (archived at https://perma.cc/E8T9-L2N2)

46 CLIA (2023) cruising.org/-/media/CLIA-Media/StratCom/Charting-the-Future-of-Sustainable-Cruise-Travel_October-2023_11-Oct-2023 (archived at https://perma.cc/J6PJ-P4XN)

47 Ibid

48 Training Aid/Martin Heng (2023) trainingaid.org/ideas-and-insights/accessible-tourism-solutions-businesses-and-destinations#:~:text=With%20the%20growing%20importance%20of,strong%20economic%20imperative%20to%20be (archived at https://perma.cc/6B89-J3UG)

49 We The 15 (2024) www.wethe15.org/ (archived at https://perma.cc/FUC2-RT58)

50 Crescent Rating (2024) www.crescentrating.com/download/thankyou.html?file=iycPeNNH_GMTI_2023_Report_-_Final_Version_-_1st_June.pdf (archived at https://perma.cc/P4PC-NZY9)

51 Ibid

52 Ibid

53 GBTA (2024) www.gbta.org/wp-content/uploads/GBTA-BTI-2023_Executive-Summary-FINAL.pdf (archived at https://perma.cc/84NT-6EYK)

54 UK Government (2023) www.gov.uk/government/publications/greenhouse-gas-reporting-conversion-factors-2023 (archived at https://perma.cc/5LQD-X8PE)

55 GBTA (2024) www.gbta.org/business-travel-industry-anticipates-a-strong-but-challenging-2024-according-to-latest-gbta-poll/ (archived at https://perma.cc/EC45-EYVB)

56 GBTA (2024) www.gbta.org/global-business-travel-industry-forecast-is-for-accelerated-rebound-spending-to-reach-1-8-trillion-by-2027/ (archived at https://perma.cc/FZ37-BX2J)

04

Digital solutions for personalized and seamless travel

Introduction

Travel is highly digitalized, disrupted over several decades by technology with various stages of transformation. Digitalization over the years includes automated bookings by Global Distribution Systems (GDS), the rise of online travel agents (OTAs), low-cost carriers (LCCs), travel search, metasearch, contactless payments and mobile travel bookings. Technology stacks for travel businesses suffer from legacy systems and are constantly evolving. Digitalization and automation are constant as travel businesses seek to drive cost and operational efficiencies while consumers increasingly embrace digital across the customer journey for ease, convenience and seamlessness. Traditional artificial intelligence (AI) has been integral in online travel for delivering more personalization to tailor search results, recommendations, marketing and offers to consumers. Travel companies have taken a leaf out of Amazon and Netflix's personalization strategies by using big data and algorithms.

The travel industry is beginning a new phase of disruption with the advent of Gen AI. Already many travellers are interested in using Gen AI such as ChatGPT to find their perfect trip. Travel businesses are creating their own Gen AI solutions or working with players like OpenAI and Google. Travel businesses are falling over themselves not to miss out by deploying Gen AI solutions for trip planning, personalized recommendations and bookings. Gen AI is just at the beginning of its journey and provides a new tool to help solve the world's greatest challenges such as climate change and poverty.

Digital technology is predominantly used for the inspiration, discovery, search and booking stages of the customer journey. The amount of time that consumers spend planning for the perfect trip is staggering. For example, US consumers take on average 45 days and almost 5 hours of search and discovery of multiple online sources and websites before booking.[1] Online booking is forecast to reach 73 per cent of travel sales by 2029 and is the primary booking channel of choice for travellers (see Figure 4.1).[2] The pandemic was a major accelerator for digital travel adoption.

There are already big moves into the latter stages of customer journeys such as the in-trip and in-destination stages for a truly personalized connected trip. The seamless and personalized travel experience is the holy grail for travel businesses and destinations. The G20 even goes so far as to link safe, seamless and enhanced traveller experiences with half of the UN SDGs for sustainable development. Where tourism is a driver for good, seamless travel plays a role in creating a win-win for consumers and travel businesses, using tech to be impact-conscious.

Already the world is experiencing Industry 4.0, the fourth industrial revolution, thanks to advanced data analytics, connectivity, automation, human-machine interaction and computer power. A new era of hyper-connectivity will be ushered in with the next generation of the internet, 6G from 2030. Wireless connectivity will be at least 100 times faster than 5G enabling intense communications anywhere, anytime, with anyone and anything.[3] Hyper-connectivity will be enabled by the IoT, edge computing and AI with ever greater internet speeds and mass generation of data. Countries like South Korea and China are leading the race to 6G.

Smart wearables are evolving to include smart contact lenses, smart glasses and even connected wireless implants in the future. Google, Meta, Apple and Samsung are already fighting for pole position in the next generation of smart wearables using AR/VR and MR such as Apple Vision Pro in spatial computing. However, the failure of Google Glass smart glasses shows that the path to success does not always run smooth.

New hyper-connected devices will enable travellers to engage in new frictionless and contextual ways with destinations and communities. The issue is whether the new hyper-connected travel experience will create an even greater divide between visitors and communities, further exacerbating

FIGURE 4.1 World travel sales by booking channel 2024–2029

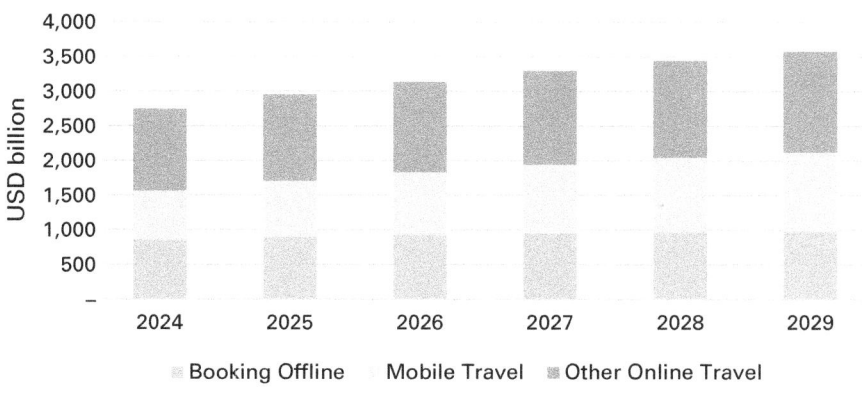

SOURCE Euromonitor International, Travel 2025 edition
NOTE Booking includes leisure travel and business travel categories

tensions with the haves and have-nots. Technology can also democratize access to products and services, lifting up communities through co-creation, new jobs and boosting income generation.

Streamlining the customer journey

The growth in internet and smartphones has powered the shift to online travel bookings. In 2000, there were only 12 mobile cellular subscriptions per 100 people, and in 2022 there were more mobile subscriptions than people at 108 per 100 people.[4] This rate of growth is phenomenal and has transformed how consumers search, book and experience travel.

The travel marketing and sales funnel is highly complex, starting with the inspiration and discovery stages. This wide part of the funnel is highly fragmented with a plethora of brands trying to capture attention. Included are a variety of different touchpoints from social media sites like Facebook, Instagram, TikTok, as well as influencers and bloggers, to search engines like Google and Bing, review sites like Tripadvisor, and metasearch sites like Kayak aiming to convert lookers into bookers. Above-the-line media like print and TV continue to exert influence even in the digital age. Yet it is the human touch that still wins out with the majority of consumers (55 per cent) talking to friends and family in-person for travel tips and recommendations.[5]

REAL-WORLD EXAMPLE

Google – kingpin in travel yet complex role in the travel ecosystem

Google is by far one of the most significant companies operating in travel, even though it sits outside of travel as an advertising company. It is by far the dominant search engine accounting for over 80 per cent of desktop searches and 96 per cent of mobile searches.[6] Its search engine and Google Ads are used by the majority of travel businesses to generate leads, referrals and ultimately convert into sales.

It has a complex relationship with the travel industry. On the one hand it receives billions of dollars from global tech giants such as Booking Holdings, Expedia, Trip. com and Airbnb. For example, combined they spent US$16.8 billion on sales and marketing that includes spending on Google advertising in 2023.[7] On the other hand, Google competes directly with travel businesses by offering Google Travel products like flights, hotels, experiences and short-term rentals. Its search listings also prioritize direct suppliers like hotel players such as Accor, Hilton, IHG and Marriott so there is a big tension when it comes to Google's impartiality. For this reason, Google, along with Apple and Meta are under investigation by the EU.

Google is deploying Generative AI in Google Search if consumers have selected Search Generative Experience in the US, which is powered by Gemini, Google's multi-modal AI. Based on the user's criteria, itineraries with activities throughout the day are provided for destinations based on web results from 200 million business profiles that Google has captured.[8] For example, 'plan me a trip to New York that includes art galleries and shopping'. Google Multisearch uses AI to understand the real world and is positioned as having your own personal tour guide. It is this overlay of the digital on the physical leveraging of AI that delivers a more personalized experience that is more contextual in real-time. Further updates to Gemini Advanced were announced in May 2024, taking personalized and customized trip planning to a new level. AI delivers complex itineraries, taking account of space-time scheduling, connectivity while integrated with Google Maps and Gmail.

Google Maps and Google Translate are also important tools for travellers. An important new enhancement in Google Maps is the provision of lower carbon alternatives to driving and flying. Alternative public transport and walking functionality is being rolled out in key cities around the world including Barcelona, London, Paris, Montreal, Amsterdam, Sydney and Rome. Long-distance rail is being introduced as no-fly options in 38 countries and long-distance bus in 15 countries, highlighted as 'climate friendly'. Rail alternatives will also be available in Google Flights. Google's open-source Travel Impact Model (TIM) is used to provide carbon emission details for flights to help consumers make a more sustainable choice.

Gen AI for travel planning

Generative AI made the headlines with the launch of ChatGPT from OpenAI, delivering this multi-modal large language model (LLM) free to the public. These LLMs are trained on vast amounts of text and images, able to understand it and generate new content. Consumers and businesses rushed to adopt this new tool with its conversational style where ChatGPT broke new records as the fastest growing app, reaching over 1 million monthly active users in three months.[9] LLMs and Generative AI allow for a freer, more natural interaction with users.

For travel businesses, this marked a new dawn in automation with certain tasks such as content creation, coding and language translation moving from people to AI. For travel businesses, there was a sense of being left behind if not jumping on Generative AI's capabilities to integrate its functionality into websites and mobile apps for personalized user experiences, while traditional travel agencies feared that Generative AI would make their services redundant, leading to a new era of closures. Technology players like Apple are partnering with OpenAI to integrate ChatGTP functionality such as trip planning into their AI assistant Siri on devices for free as the AI race heats up.

AI is not a new priority, it has been a significant investment priority for travel and mobility businesses for the past five years, accounting for 65 per cent, versus 13 per cent for IoT, 10 per cent immersive tech, 7 per cent web 3.0, 3 per cent 3D printing and 2 per cent for quantum computing.[10]

Companies like Kayak and Expedia were among the first businesses to integrate ChatGPT functionality through plug-ins to provide more tailored and personalized trip planning recommendations in a conversational way. In May 2024, Expedia went one step further by introducing its AI Travel Assistant, Romie, designed to help throughout the entire trip from planning to in-destination through hyper-personalization. TUI offers ChatGPT functionality as part of its mobile app, while TripAdvisor's Generative AI planning tool even delivered incremental revenues with members that saved an AI-generated itinerary, delivering three times more revenue than those members that did not.[11]

REAL-WORLD EXAMPLE
Trip.com – the world's leading OTA adopts Gen AI to deliver the ultimate user experience

Trip.com Group is the largest global OTA thanks to its dominance in the large Chinese domestic and outbound tourism markets. Gen AI has breathed new life into the company after China's reopening post-pandemic, refreshed by a new spirit of 'innovationism'.[12] Trip.com launched its Gen AI assistant, TripGenie, in the app for travel planning and booking for a hassle-free and personalized travel experience in February 2023. For Trip.com, the LLM delivers customized results, seamlessly leading from recommendations for hotels, transport and experiences to booking links.

TripGenie is present throughout the journey providing contextual assistance. The AI assistant refines information based on user preferences, removing the need for filtering, making travel planning and booking as natural as a chat conversation. The tool is also highly collaborative so that multiple users can create tours and itineraries together, making group travel much easier. With real-time updates during the trip, AI helps to navigate changing circumstances such as delays or weather. Future development includes real-time voice translation.

AI is not just about delivering for consumers, but a revenue driver. The introduction of Trip.com's AI assistant has increased bookings by 90 per cent and led to more time spent on the app.[13] Rolled out across 200 countries, AI has helped to boost repeat visitors, strengthening brand loyalty – all part of creating a resilient business model.

With a younger consumer base of Millennials and Gen Z that are digital natives, Gen AI is a logical next step to provide a more personalized and social-media savvy service that leverages user generated content (UGC), reviews and peer recommendations. Innovation like TripGenie is just one part of a broader toolkit for customer service delivery, part of a sophisticated ecosystem built from cutting-edge technology stacks and people working on innovation, R&D, customer service and marketing.

Other functions for AI in travel businesses besides customer service include predictive analytics for optimization of resources, supply and demand, robotics and autonomous vehicles. One of the most compelling features of LLMs is the translation capabilities, which is often one of the biggest challenges that people face when travelling abroad. With LLMs and their instant translation, this is no longer a barrier between visitor and local communities. In South Korea, AI-powered instantaneous translation is provided in select Metro stations where a transparent screen between staff and visitors displays the translated spoken information in 11 different languages.

Gen AI is increasingly the new normal for travel businesses in transforming to a sustainable future, where 33 per cent of travel sellers saw great opportunity to drive personalization and sustainable travel behaviours across the traveller journey.[14]

Ubiquitous AI requires a responsible approach

Consumers are embracing AI across all aspects of their daily lives, whether from OpenAI, Google or Microsoft, for different purposes such as work, banking, social media, life hacks, romance and healthcare. There are high expectations that travel businesses also provide similar if not even more elevated and hyper-personalized services. Generative AI and traditional AI are pervasive, and it is not always visible where they are being used.

The dangers and risks of AI are also being voiced not only from the sidelines but from its very creators. OpenAI's CEO Sam Altman has been vocal about the threats from AI if governments around the world fail to regulate its development. Altman has warned of AI's existential threat to humanity, even mentioning the end of the world, and subtle societal misalignments.[15] The primary areas of concern regarding threats posed by AI are safety, security and trust in terms of providing disinformation, and promoting discrimination and bias.

The EU is the first to introduce an AI Act to address concerns as well as deliver the benefits of AI. The technology is touted as a means to combat climate change and help to deliver the net zero transition agenda in areas such as transport and clean energy. The EU advocates for AI to be managed in a safe, transparent, traceable, non-discriminatory and environmentally friendly way, under the oversight of people. The US administration introduced its Executive Order 14110 on how to manage AI safely, securely and responsibly with a bill still under discussion.

With the use of Generative AI in social media platforms like Meta AI powered by Llama 3 or enterprise software such as Microsoft's Copilot for Office 365, its presence will be ever more ingrained in businesses and consumers' daily lives. Yet consumers are increasingly worried about their privacy and surveillance, while regulatory standards are still being developed.

In the US, 52 per cent of consumers responded that they are more concerned than excited about AI regarding the increased use of AI in daily life.[16] Despite the fast adoption of AI there is a low awareness of some of the risks with 73 per cent of consumers globally trusting content that is created by Gen AI.[17]

Impact of AI on travel businesses

There is a myriad of ways that AI is being adopted by travel businesses and destinations to drive efficiencies operationally and in terms of customer experience. Key uses include providing personalization by leveraging previous consumer behaviour and purchasing habits, predictive analytics for pricing and demand forecasting, the use of chatbots and voice assistants. Optimizing workforce schedules, machinery and operations across the built environment are also where AI excels. Robotics is another area where AI is present such as robot butlers, luggage bots, mop bots or robotic bartenders as seen on cruise ships like Royal Caribbean's Bionic Bar or MSC Cruises among others.

Machine learning and data analytics were the top investment priorities for travel businesses, in 2024, while machine learning and Generative AI will be the top priorities in 2029 according to Amadeus.[18]

ROBOT CHECK-IN

Emirates set a world-first with its robotic check-in assistant at its downtown City Check-in and Travel Store in Dubai's financial hub. The Emirates robotic assistant complements self check-in and advisers, and conducts passport biometric ID checks, flight and baggage check-in. With the roll out of robots for check-in, this leads to the closure of check-in desks and potential job losses or redeployment.

TIPS

- The role of robotics in customer-facing roles is more of a gimmick in such a people-centric industry like travel, yet next generation automation poses a very real threat of job losses.
- Giving consumers the choice of interacting with a robot, human or both will ensure that different comfort levels of engaging with AI are met.
- Understanding differences to AI across the generations is key.

However, the biggest fear especially among MSMEs and travel agencies are what are the risks to jobs following this latest leapfrog in automation. Tech titans including Sam Altman and Elon Musk predict the loss of jobs; in the latter's opinion all jobs are at risk due to automation, leading to the need for universal basic income. On one end of the scale, there is no reason for alarm. Technology experts like Gartner predict a neutral impact by 2026 with no net gains or losses, with 0.5 billion new jobs created as a result of AI solutions by 2033.[19]

The level of uncertainty and disruption is reason for concern. Potentially 300 million jobs could be exposed to automation from Gen AI, which is a boon for the global economy of US$7 trillion or 7 per cent over 10 years due to productivity gains.[20] This requires travel business leaders to get ahead of the challenge to the labour force, future-proofing by harnessing AI, while focusing on upskilling and reskilling works for an AI-first future in the decades to come.

While Gen AI can be deployed to deliver greater efficiencies of resources, its computational power is more carbon intensive, requiring a 'both/and' approach to sustainability and digitalization. It is necessary to take a holistic view of environmental impacts to mitigate its higher carbon footprint, bearing in mind the data centres can switch to renewables and use energy-efficient cooling technology.

A clear advantage of human-centric travel businesses is the ability to oversee AI and ensure that it does not produce false information. The role of people to observe, vet and intervene to prevent machine hallucinations that create false and misleading information is a fundamental USP that will help to keep travel businesses viable and sustainable.

'Today AI is the worst it will ever be – it only gets better and more powerful from here. But sadly adoption of AI is currently much lower in travel and tourism than other consumer facing industries, so travel businesses should start using it now. It can be used for big challenges like decarbonization, to maximize business efficiencies and in office settings to drive creativity in departments such as marketing.

Business leaders and staff at all levels of an organization also need to urgently start learning how to use it otherwise the travel and tourism sector will be left behind and seen as a dinosaur industry, especially in the search for talent when AI native kids who have grown up with the technology come into the workforce in the future.'

James McDonald, Director, Travel Transformation, World Travel and Tourism Council

---

TIPS

- Despite the AI race, consumer privacy, safety and security must be paramount.

- Currently, AI-generated search results are presented to consumers for them to decide what is the best option, taking into account climate impacts, convenience and price.

- In future, travel businesses may intervene to block or stop non-climate friendly options provided by AI depending on future regulation.

- Alarm bells are already sounding regarding AI and its potential to wreak havoc on societies. A measured and responsible approach is required to be led by government.

- Transparency on where and how AI is used especially regarding content and information is important to build trust.

- Climate impacts should be considered when making the move to Gen AI.

- Human-centricity as opposed to automation in travel is a strong USP for social sustainability.

- Gen AI will never replace the human and emotional connection between people.

Major concerns regarding all AI are that they can perpetuate biases from the web sources referenced. However, taking an active engagement to use AI for a purpose such as DEI means that checks and balances are in place to help prevent bias creep.

Emotional AI that can interpret and understand human emotions using facial recognition, for example, has the potential to deliver even more personalized responses. Yet, the technology has been parked by major companies like Microsoft due to ethical concerns.

There are ongoing concerns about the digital divide that AI and digitalization causes, creating exclusion for certain groups such as seniors. Digital exclusion should be taken into account to ensure that the technology is safe and open for all to access.

AI garners investor attention

Although travel is one of the slower industries to adopt AI, AI has been a central tenet to the most successful travel tech start-ups in the past several years. Innovative travel tech players like Hopper with its predictive pricing and flexibility including cancellations of non-refundable flights have raised millions in investment. Its B2B division Hopper Tech Solutions (HTS) is gaining traction, rolled out initially with CapitalOne Bank (US) and providing a white label travel marketplace for hotel and resorts for its Capital One Venture X card members. It has strategic partnerships with Uber and Kayak, Trip.com and Marriott for reselling its Cloud services.

Some start-ups like Layla are virtual travel agents, based on an AI persona that provides Instagram videos for personalized trip itineraries powered by Booking.com, Skyscanner, Beautiful Destinations and Get Your Guide. Layla acquired Roam Around in 2024. Others like Guide Geek combine human curation from travel publisher Matador with Gen AI to integrate into social media platforms like Instagram, WhatsApp and Messenger for free.

TIPS

- Adoption of AI should be responsible and avoid exacerbating the digital divide.
- AI opens up opportunities for start-ups to solve the last-mile challenges and reduce the expectation/experience gaps that persist throughout the traveller journey.
- AI-first travel start-ups need to ensure that unsustainable tourism practices are not perpetuated, so should avoid promoting mass tourism and bucket-list travel.
- AI applied to positive impact tourism can help to drive travel businesses and consumers towards a better outcome for destinations, communities and visitors.

'For now, Generative AI is multi-modal, acting as an enabler rather than a replicator with businesses adopting for automation, travel recommendations, emotional recognition, customer service, translation, robots, holograms, avatar creation. Advanced uses include Google Gemini AI for maps, flights and complex itineraries or companies like Obvlo providing curation and personalization of experiences. However, it is not yet real-time so faces challenges creating plans that are adaptable for weather or travel disruption.'

Joshua Ryan-Saha, Director, TravelTech for Scotland and Edinburgh Futures Institute

AR/VR go mainstream, with further evolution to MR/XR

Part of trip planning is the uncertainty of what awaits the traveller at each stage of the journey. Virtual reality (VR) is not a new digital tool, immersing the user fully in a digital world, while augmented reality (AR) overlays digital elements on the real world.[21] Its primary purpose is to produce creative content marketing for inspiration and discovery. Accelerated during the pandemic, virtual travel experiences have become mainstream thanks to VR headsets and AR/VR apps, enabling potential visitors to try before they buy. For travellers, especially those with accessibility needs, it is an integral tool to ensuring important details are provided so that they can confidently make a booking with greater understanding of what the service provision will be thanks to 360° visualization.

Consumers can relive the past in an immersive way thanks to VR/AR headsets to enjoy Pompei, Ancient Rome or Ancient Greece in-situ, while virtual access to remote locations such as the moon, the depths of the oceans, and poles are within (virtual) reach, even fast-forwarding to the future, opening up new revenue opportunities for travel businesses.

EXPERIENCE ANTARCTICA BY VR/AR

As the most remote region in the world, only the lucky few will experience the beauty and vast expanse of Antarctica. However, it is a vital lifeline of the planetary system and at risk of massive climate change impacts. In order to engage with new and younger audiences, the UK Antarctica Heritage Trust created Immersive Antarctica in virtual reality in 2023 followed by Wearing Antarctica in augmented reality (AR) in 2024. In AR, users are invited to try six immersive experiences such as dog sledding or undertaking scientific research and conservation. The AR experience is available on TikTok and Instagram channels where people can share their stories, and most importantly engage with stories about the past and our shared future.

As the path to net zero accelerates towards 2050, virtual experiences are likely to play an ever greater role as an alternative to real-world adventures unshackled from the potential obstacles of costs, visitor bans and quotas. For example, virtual travel to Mount Everest could be a potential revenue stream for local communities if expedition treks and permits are further restricted, or even banned due to environmental damage from pollution.

Immersive experiences using AR/VR are already common in museums around the world to bring objects and history to life such as the Smithsonian Institution's AR tour or the Holocaust Museum. Theme parks such as Universal Japan's Mario Kart Koopa's Challenge, overlay AR on the real-world rollercoaster ride for a truly interactive experience.

Snapchat is the leader in AR, popular with younger generations. It works with tourism brands for marketing and brand awareness to boost engagement, while also bringing iconic monuments to life as seen with the Barbie movie AR experience turning the Eiffel Tower pink. Snapchat users are using AR throughout their travel journey including 44 per cent at the time of booking and 52 per cent in-destination.[22]

RED BULL EMBRACES MIXED REALITY

Red Bull's immersive exhibition, Water – Breaking the Surface, at the Museum of Transport in Lucerne (Switzerland) enables visitors to experience through mixed reality the adrenalin rush of cliff diving. The technology used is Varjo's MR headset along with VR, giving unprecedented access to the experience of a Red Bull cliff diver.

The use of extended reality has not fully taken off. If XR is to replace travel and become a viable alternative to travel as destinations become increasingly at risk from climate change or overtourism, the next generation of XR experiences will be required. This entails leveraging web 3.0 technology as provided by game engines like Unreal Engine owned by Epic Games and Unity that are at the forefront of real-time 3D experiences.

The role of AR/VR will increasingly blend the virtual and physical worlds for 'phygital' seamless and immersive travel experiences. The evolution into mixed reality (MR) takes AR into a more interactive domain where consumers can engage and interact with digital objects and information, while extended reality (XR) encompasses all aspects of AR/VR/MR as well as what is still to come in future. The immersive, intelligent and connected experience from XR smart devices will take a while to become mainstream with challenges to overcome including perception, visualization, power, heat, edge computing and connectivity. Apple Vision Pro, positioned as spatial computing, is too cost-prohibitive for it to be widely adopted.

The next generation of affordable smart XR devices will lead to hyper-connected travellers, with the immense power of the web to guide them.

This may lead to further disconnects between visitors and local communities, as well as travellers themselves caught up in their own phygital worlds. A potential benefit is greater awareness of personal impacts through lifestyle carbon footprint tracking apps on such devices.

Web 3.0 future developments

The next generation of internet, web 3.0, will be more immersive and interactive, built on the technology of today, however, not everything is known about how it will look in the future. Web 3.0 is characterized by open-source decentralized networks, powered by blockchain cryptocurrencies, offering greater personal control of data and agency. One manifestation of the next generation internet is the metaverse, simultaneously derided and embraced by many travel businesses, mainly due to hype and fear of missing out (FOMO).

The metaverse is defined as a network of connected virtual worlds, created by the convergence of virtually enhanced physical and digital reality, where the experience is persistent and immersive.[23] Essentially, it's where gaming meets work, play, travel and socializing. It leverages decentralized technology like blockchain cryptocurrencies, non-fungible tokens (NFTs) and uses XR headsets, AI, IoT and cloud technology.

Some destinations like Seoul launched a metaverse platform for its residents including a virtual tourism zone for visiting top attractions. An important pillar of the Seoul platform is inclusion and users can interact as avatars, where they can express themselves freely without fear of the discrimination that they potentially face in real life.

Barbados opened a virtual embassy in the decentralized metaverse platform, Decentraland, providing e-visas and digital cultural experiences. Some travel businesses have experimented with buying virtual land and recreating their brand experiences in the metaverse, such as hotels like Marriott and Millennium & Copthorne. Marriott launched Marriott Bonvoy Land in Fortnite, introducing its brands' experience to a newer and younger global audience through gamification in the virtual world with offers for stays in the real world. It was the first global hotel brand to make a foray into web 3.0 in 2021, with a drop of NFTs by digital artists at Art Basel Miami Beach (US).

TripAdvisor is reportedly taking a leap into the metaverse so that its consumers can experience sites and attractions virtually beforehand for a more interactive engagement with the brand and destination. Although, with financial challenges and talks of a merger, it remains to be seen whether this initiative will fly.

Changi Airport (Singapore) is the first airport in the metaverse with the Changiverse via online game Roblox. Here consumers can play games, discover the airport's features such as the waterfall, collect virtual butterfly rewards and swag prizes. Changi is a world-class airport in real life, where embracing web 3.0 introduces it to a younger and global audience.

The metaverse has gone from hype, thanks to Mark Zuckerberg's renaming of Facebook to Meta, to claims of its demise. The next iteration of web 3.0 will be an immersive and interactive place of work, play, discovery and exploration for consumers to engage in new creative ways with travel businesses.

TIPS

- Understanding the next generation of web 3.0 and new technologies like Gen AI, edge computing and spatial computing will be essential to keep ahead of consumer expectations and behaviours.

- Decentralized and gaming platforms are where consumers, especially younger generations, are looking for new and exciting virtual experiences with their favourite brands.

- Adoption of new digital tools as seen with ChatGPT can happen at ever increasing speeds and scale, requiring ongoing commitment to digital transformation and skills.

Digital twins drive efficiencies and seamless travel on the path to net zero

As technology like IoT and AI develops, digital twins of spaces such as airports, aircraft, cities, hotels, cruise ships and even people are created and represented in a real-time 3D digital world. This enables a more sustainable and efficient approach to optimizing operations through virtual dimensional simulations, also allowing for scenario planning and real-time interventions when things go wrong. Digital twins rely on IoT using sensors that gather data in real-time, AI machine learning and deep learning, cloud computing, automation and XR devices for engagement with the digital twin such as AR smart glasses. Tech companies offering digital twin platforms as a service include Microsoft Azure, IBM, Siemens, Amazon Web Services (AWS) as well as Unity.

Real-time data from multiple sources including in-person interactions, video, cameras and IoT sensors can be overlaid on the 3D geospatial images of the asset modelled to create a spatially aware visualization of operations. As this is a model, it leverages historic data, combines with real-time data and can use predictive analytics. Applications for digital twins include check-in/check-out during the customer journey, equipment management, aircraft turnaround, passenger flow management curb to gate, and optimizing resources. Leveraging such technologies will help travel businesses that run physical assets and workforce, optimize their operations and ensure that they remain on track to meet net zero targets by 2030.

REAL-WORLD EXAMPLE

Vancouver International Airport partners with Unity to transform its operations

Unity is known as one of the world's leading game engines, founded in Denmark in 2004. As part of its professional services, Accelerate Solutions, it provides real-time 3D digital twins. Working with Vancouver International Airport, both are pushing the envelope when it comes to next generation airport operations using a people-first technology approach for its 25 million passengers and workforce.[24] Besides driving efficiencies in the airfield and terminal, the digital twin is aligned to deliver on climate and reconciliation goals promoting the indigenous community's, Musqueam, participation.

In line with the digital twin concept, Unity has recreated Vancouver International Airport in a 2D or 3D way, using historical and real-time data from IoT for optimization, planning, training and scenario planning. The airport is working with Unity and airlines to build a carbon emissions calculator, to measure and visualize emissions from aircraft take-off and landing movements to reduce airside emissions.

The digital twin acts as a 24/7 situational awareness tool so that staff can take necessary or pre-emptive measures for passenger safety and seamless customer experiences. This means that decision-making is more informed and timely. Unity and the airport signed a memorandum of understanding (MOU) to expand the digital twin for airports concept globally and boost digital and sustainable transformation in the sector. Game engines like Unity are at the forefront of new technologies that blend the virtual and physical worlds to improve travel experiences for all.

Facing existential challenges, destinations turn to digital twins

Cities like Helsinki have invested in a digital twin since the 1980s driven by the desire to be more sustainable and resilient. Countries like Singapore are among the countries leading the charge to become the ultimate smart nation, aiming to be digital first across government, society and the economy. Singapore's digital twin, Integrated Environmental Modeller, visualizes metrics such as natural ventilation, noise, sun irradiance and air temperature to test urban planning solutions and understand the cost-benefits before moving ahead with new plans.

Cities across Europe are embracing digital twins to future-proof their resilience by modelling future scenarios from Barcelona, Flanders and Rotterdam. The Barcelona model is built on the 15 minutes model where services should be within reach within that time.[25] It can assess the level of social inclusion or exclusion as well as environmental and impacts. Further interoperability and connecting of Local Digital Twins are expected in Europe creating the CitiVERSE – the EU's alternative to the metaverse – funded by the European Digital Infrastructure Consortium (EDIC).

In the case of Ukraine, digital mapping of monuments already destroyed or at risk of bombing in the Ukraine/Russia war are used to preserve the country's cultural heritage for future generations in the virtual world.

REAL-WORLD EXAMPLE
Tuvalu builds a virtual future in response to existential climate threat

The plight of Tuvalu, a Pacific Island nation, that faces existential destruction from climate change, went viral at COP26. Minister for Justice, Communication and Foreign Affairs, Hon Simon Kofe, addressed the COP from a podium that transpired to be surrounded by sea water, highlighting the country's threat of extinction from rising sea levels, natural disasters, droughts and ocean acidification. Due to its low-lying position, it is at the forefront of climate change and faces 50-year floods every five years by 2060.[26] Human-caused climate change is the most serious threat to the country's existence and its right to sustainable development.

By building a digital twin leveraging web 3.0, Tuvalu aims to preserve its national identity, culture and sovereignty. At COP27, Kofe introduced Tuvalu as the first digital nation, speaking from the metaverse in response to the disappearance of their home.

In partnership with Accenture and Unreal Engine 5, Tuvalu recreated one of the first islands expected to be submerged in the metaverse, Te Afualiku, as part of the Future Now Project. If the population needs to relocate then this plan ensures that the country will continue to exist digitally, providing administrative services and maintaining its statehood, international voting rights and marine boundaries. The move into an immersive virtual world may seem radical yet it highlights the extreme challenges faced by small island developing nations.

Businesses and MICE move into immersive digital worlds

Despite the term metaverse having been stigmatized, there is a move by businesses from fashion to retailers into the immersive web. Start-ups such as Journee in Europe are leading the charge, working with iconic brands across fashion and luxury like Macys and H&M. It offers tourism solutions for inspiration and discovery, pre-trip and post-trip including virtual events and virtual check-in, personalized recommendations and AR/VR tours. Other web 3.0 start-ups are attracting attention from meetings/incentives/conventions and exhibitions (MICE) and hospitality players. Rendezverse, originally set up as an alternative to in-person events, has pivoted to providing a venue search solution, using AI to match buyers with event properties including some of the world's most famous hotel groups like Marriott and Hilton. The demise of MICE has not come about, and if anything web 3.0 and AR/VR are being leveraged to create more interactive and engaging experiences in real life.

REAL-WORLD EXAMPLE

Airbnb – from digital disruptor to global lodging leader

Love it or loathe it, Airbnb has transformed the travel landscape in the past two decades and continues to spearhead digital innovation in its services and customer experiences with the adoption of Gen AI.

Founded in 2008, it has enjoyed phenomenal success by digitalizing the fragmented short-term rentals sector – with the company synonymous with holiday rentals and the 'live like a local' trend. The company is present in over 220 countries and regions worldwide, with 5 million hosts and welcoming over 1.5 billion guests. In terms of gross booking value, the company recorded US$73 billion in 2023, up 14 per cent on 2022. Profit was the star performer in 2023, up by an impressive 153 per cent.[27]

Its success lies in meeting the different needs of its dual hosts' and guests' communities on its platform. The former are looking to monetize their physical assets, leveraging Airbnb's wealth of data on traveller preferences to drive personalized search results, marketing, pricing and customer service, while consumers embrace the seamless online booking platform, flexibility and choice of Airbnb's offer from off-the-beaten-track locations to urban centres. AI, machine learning and automation increasingly play a fundamental role in delivering for its communities.

Airbnb's CEO Brian Chesky has big plans for Gen AI to be the ultimate travel concierge, knowing guests and personalizing services to anticipate guests' needs, even moving beyond travel into lifestyle services. Airbnb's acquisition of Gameplanner.AI takes the company one step closer to achieving its goals of enhanced human connection through AI. Although the company is not going to build its own AI models, it will continue to work with technology partners like OpenAI,

Google and Meta to build its own AI interfaces to drive ever greater customization. Group booking functionality using AI was added in 2024 to further provide greater levels of personalization.

Yet, Airbnb's success has not been plain sailing. Challenges include badly behaved guests, safety and security concerns, fake listings, fraud and regulatory battles once legislators caught up with the proliferation of residential properties converted to short-term lets. Protests in key overtourism hotspots like the Canary Islands draw negative attention to Airbnb's contribution to the housing crisis. Adopting the latest technology such as AI helps Airbnb to tackle fraud, however, AI cannot solve all of Airbnb's problems.

Seamless travel – safe, secure and touchless to speed up the journey

There are multiple ways to interpret a seamless travel experience. Ultimately it is about a person travelling safely and securely in terms of sharing their personal information, in the most smooth and efficient way, where digitalization speeds up the process, making it contactless. This smooth end-to-end customer journey from booking, transport check-in, hotel check-in and check-out is enabled through touchless technology, biometrics and interoperability. Given that travel and tourism are identified as a means to drive positive impacts for communities, making travel seamless contributes to the SDG agenda.

For travel businesses, the customer journey end-to-end is a complex chain of handovers where there is much scope for friction and for things to go wrong. This complex supply chain is often at the mercy of events outside of the businesses' control such as geopolitics, strikes, weather or force majeure – other acts of God. Safety and security are inherent to any trip. The attitudes, perception and level of risk that each individual faces are different. However, once those factors have been addressed and travellers are aware, the next priority in the hierarchy of needs is convenience before ultimately enjoying themselves and fulfilling their aspirations.

Over the past decades, the travel industry globally has introduced many protocols to facilitate travel. Travel authorization is fundamental, as determined by nationality and associated visa agreements. Globally, 49 per cent of consumers are deterred from travelling abroad due to complex visa processes.[28] Bilateral visa agreements come and go so this can often be a minefield for travellers, especially in the case of the UK, following Brexit and leaving the EU to become a third country. Even in the digital age, pain points continue to persist and challenge travellers including baggage drop-off and pick-up and queues at security and border control (see Figure 4.2).

FIGURE 4.2 Consumers' most dissatisfied touchpoints in the air traveller journey

Most Dissatisfied Touchpoints in the Airport Experience 2023, ranked by the largest percentage

| Baggage collection | Border control / immigration | Security | Baggage drop | Boarding | On-board | Transfer | Check-in | Search for travel option | Airport arrival | Booking | Final destination reach |

SOURCE IATA – IATA Global Passenger Survey Highlights 2023[29]

Digital traveller identity

The digital traveller identity is a very important pillar of seamless travel. Security has seen a major overhaul thanks to biometrics and AI, speeding up the journey at borders including ports, airports and railway stations. Biometrics is used to verify a traveller's identity based on their passport or mobile phone or through identification with an associated reference number in a biometric enrolment database.

Consumers expect the traveller journey to be even faster and the vast majority (87 per cent) are willing to trade their personal data to enjoy a smoother and faster airport experience.[30] Biometrics are increasingly preferred to boarding passes or passports.

In terms of airports that are getting it right when it comes to customer satisfaction, Rome Fiumicino won the award for best security screening experience, while overall Doha Hamad was ranked top airport in 2024.[31] In terms of immigration service, Changi Airport (Singapore) was rated top, followed by Zurich and Bahrain Airports.[32]

Eurostar introduced fast-track contactless kiosks powered by biometrics to facilitate and speed up the check-in process. Eurostar has partnered with Iproov and Entrust for its SmartCheck solution that follows the ICAO Digital Travel Credential (DTC) for a touchless check-in and border processing experience.

The path to digital ID does not run smoothly. There is concern from governments and civil liberty groups about excessive surveillance using facial recognition and biometrics. In the US, there is push back on the paid-for membership scheme for TSA PreCheck or CLEAR Plus roll out. The scheme allows travellers to pass quickly through security in a less intrusive way with no need to remove shoes, laptops or liquids, available in 200 airports in the US across 90 airlines.

Trusted traveller programmes are increasingly used to expedite travellers through travel hubs. The US's Global Entry system is being emulated in regions like Europe. Global Entry aims to move low-risk travellers quickly through the customs and border process based on pre-approval clearance. Global Entry is also available in Toronto and Vancouver (Canada), Dublin and Shannon (Ireland), the Bahamas, Aruba and Abu Dhabi (UAE) among other international locations. However, despite the success of such schemes, the US sits in 17th position out of 18 countries assessed for their proactive and seamless travel facilitation.[33]

In Europe, the introduction of the European Travel Information and Authorization System (ETIAS) in 2025 is preceded by the Entry/Exit Scheme in 2024. Countries like Finland are moving first in terms of the use of the EU Digital ID Wallet (EUDI) and working towards interoperability for migration, border control, law enforcement and security across all member states.

The WEF launched the Known Traveller Digital Identity in 2016 and work continues on making crossing borders more seamless. IATA is testing its One ID programme that is passenger-centric across the entire end-to-end customer journey across multiple stakeholders. The first trial was a British Airways flight from London to Rome Fiumicino in 2023. Only 17 per cent of airlines have introduced single token ID, while 38 per cent plan to by 2026 so the ability to show ID only once throughout the passenger journey is still a long way off.[34]

By 2026, 70 per cent of airlines expect to have biometric ID management installed with 90 per cent investing in a touchless passenger experience and reducing curb congestion.[35] As digitalization continues apace, the priority for ensuring cybersecurity becomes even more critical. The global IT outage caused by CrowdStrike in 2024 reinforces the need to accelerate cybersecurity investment and develop strong resilience strategies.

Traveller ID – single digital ID token

The first example of the biometric single token ID was by Aruba and the Netherlands in partnership with Visionbox, KLM and Aeropuerto Internacional Reina Beatrix, a truly cross-sector collaboration. Named Aurba Happy Flow, this trailblazing solution is the blueprint for providing a seamless end-to-end customer journey while maintaining security. Moving to a single identity check using a biometric token removes a lot of pain points, speeding up the process for passengers, while different stakeholders have access to a shared IT platform with privacy baked into the design and primed for interoperability. Visionbox was acquired by one of the world's leading GDS players, Amadeus, in 2024 aiming to provide a seamless and integrated customer experience from booking to boarding.

IATA's One ID concept is based on a decentralized ID that is verified without the need for a centralized verification, which enables consumers to take control of their own data. This marks a shift towards decentralization which is where the next generation of the internet is heading with web 3.0 powered by blockchain and decentralized commerce. This shift is away from the current central control of data by tech giants and travel businesses.

TIPS

- Biometrics is a controversial topic yet consumers tend to comply if they feel that their personal data is safe and secure.

- As consumers increasingly take more control of their data, they will have higher expectations about the speed, ease and seamlessness of their customer experience as a trade-off.

- There are no shortcuts when it comes to data privacy, consumer protection and cybersecurity.

- Trust is absolutely critical and should be the primary concern of travel businesses.

- Developing cyber-resilience is a fundamental pillar of dealing with potential cyberattacks and recovering as quickly as possible with minimal damage to brand reputation.

Privacy and customer-centricity by design

The accelerated digital transformation that is ongoing in the travel industry requires continuous investment in protecting IT systems from malicious cyberattacks from hackers and faulty updates. The risks to businesses, big and small, are enormous in terms of brand image and trust, which can be destroyed with a single attack. The concern is so high amongst travel leaders that cybersecurity is deemed a greater priority (62 per cent of respondents) than sustainability (51 per cent).[36]

With leaps forward due to Gen AI, cybersecurity is demanding ever greater attention and regulations are playing catch up to ensure data privacy and security. The types of cyberattacks range from phishing, data breaches, ransomware attacks on websites, operational disruption, to social media scams. Data privacy regulations such as General Data Protection Regulation (GDPR) in Europe are being rolled out globally.

Contactless and flexible payments

Besides secure digital identification, digital payments are another fundamental piece of the puzzle in delivering a seamless and personalized traveller experience. The payments ecosystem of travel and tourism businesses is highly complex especially for cross-border transactions. There are multiple different payment platforms and solutions from online/offline, direct vs indirect, B2C, B2B, B2B2C, peer-to-peer, flexible payments, buy now pay later (BNPL), blockchain, cryptocurrency to contactless and digital wallets.

Contactless ensures secure payments as well as the ability to enjoy rewards and benefits. Almost three-quarter of travellers in the US use digital wallets to improve their travel experience, while in Asia Pacific, 97 per cent of travellers are going contactless, using credit/debit/prepaid cards or digital wallets, and only 17 per cent use cash.[37, 38]

Cash is in a descending spiral as consumers, businesses and even central banks shift to digital. One hundred and thirty-four countries accounting for 98 per cent of world GDP are considering the shift to central bank digital currency (CBCD), with Jamaica, the Bahamas and Nigeria being the first to introduce them.[39] In the UK, the Bank of England is also considering a digital currency.

With the proliferation of smartphones globally, m-commerce and digital wallets are helping to drive digital inclusion, capturing the unbanked and allowing them access to finance without having access to a bank account. This has been particularly successful in emerging markets with the rise of superapps such as WeChat and Alipay in China, Grab Pay, GoPay in Southeast Asia, Careem in the Middle East, M-PESA in Africa, while Rappi is making big inroads in Latin America.

TIPS

- Payments constantly are being disrupted and embracing digital transformation.
- Partnering with fintechs is one way to stay head of the curve.
- For travel businesses, delivering the most customer-centric and simple solution will resonate most with consumers.
- Flexibility is key.

Superapps – one-stop shop for lifestyles

Advanced regions of Europe and North America can only look on with envy as emerging regions like Asia Pacific lead the world in superapps. These are mega apps that contain mini apps within their ecosystem, enabling cross services to come together under a unified platform.

WeChat owned by Tencent in China is by far one of the most used apps in the world, with 1 billion active users and revenue of US$16 billion in 2023.[40] Starting off as a messaging app, it has added payment functionality through WeChat Pay and a host of apps like Meituan including lifestyle services such as travel, tickets, food delivery and ride-hailing. Didi is another major app on the WeChat e-commerce platform, offering Uber-style shared mobility and food delivery for smart transport solutions. However, Didi has come under scrutiny from the Chinese government and the US government.

AirAsia, a low-cost carrier, attempted a superapp but had to rethink its strategy as it did not deliver the results expected. Its app has been rebranded as AirAsia MOVE and still has the functionality for booking flights and hotels, rewards, financial services and lifestyle services like car-hail and dining.

TIPS

- Integrated mobility, foodservice, health and wellbeing, retail and travel is a sweetspot that superapps excel at tapping into especially in Asia Pacific.
- Interoperability of payments and loyalty rewards is on the rise.
- Think laterally about ancillary services and partnerships that deliver higher value for travellers.
- Partnerships with embedded finance and fintech players are critical to build long-lasting relationships with consumers.
- Integrating planet-friendly rewards and services into payment solutions will help keep consumers engaged with the net zero agenda.

Personalized impact and lifestyle services

Personal climate calculators and impact trackers are available as apps or integrated with consumer spending purchases such as Visa's Futurecard or Mastercard's partnership with Doconomy. This provides greater transparency to consumers on their carbon footprint, rewarding good behaviour and providing tips on how to improve.

For Visa, it is a way to deliver on their net zero by 2040 agenda. Visa research found that public transport would rise by 25 per cent if it were easier to pay for, hence it's working with urban mobility providers and EV charging manufacturers to drive interoperable and seamless payments.[41]

Eco bundles, carbon calculators and automatic carbon offsetting help drive sustainable behaviours thanks to rewards. The Futurecard for example gives 5 per cent cashback on sustainable purchases like taking the bus, charging an EV, buying second hand or shopping for low carbon items.

CarbonPay is a pre-paid business expenses platform and card that provides offsetting for purchases automatically, amounting to 1kg CO_2 per GBP/US$ spent. The fintech player is powered by Visa and Stripe, is certified B Corp and part of the 1 per cent for the Planet. CarbonPay is one of Ecolytiq's first clients, a so-called Sustainability as a Service company, providing a seamless and integrated personalized carbon impact. Consumers can make a big difference to their carbon footprint by using carbon foot-print tracking apps where gamifiying positive impacts by consumers can speed up the transition to the net zero agenda.

Loyalty digital shake-up

One of the most significant shifts in the consumer landscape since the rise of e-commerce is the demise of brand loyalty, driven by a variety of factors such as too much choice, over-saturated loyalty programmes, price sensitivity and apathy, among others. That is not to say that loyalty is dead – instead it has transformed. Consumers are driven by the search for value, balancing price with the benefits of brand engagement, swayed by personalized offers and rewards.

Across the travel industry, reward programmes are prevalent, with even global OTAs entering the loyalty fray. For example, Expedia launched One Key across its brands, usurping its former Hotels.com loyalty programme, promising to reward every traveller. Its short-term rental brand, Vrbo, pipped Airbnb to the post, where Airbnb has so far continued to dodge rolling out a loyalty programme, although it talks a lot about doing so.

Airlines and hotel brands have for decades focused on their loyalty programmes and co-branded credit cards to build a closer customer relationship to drive bookings and take greater control over their inventory, away from the control of third parties like OTAs that are a necessary evil with their high commissions.

Marriott Bonvoy is the leading loyalty programme with over 200 million members, taking personalization to the next level with technology, covering 30 brands in 10,000 destinations.[42] It is deploying Gen AI search functionality for its Home & Villas offer, using natural language search to provide curated results on attributes rather than destination-specific, while Gen Z and Millennials in regions like Asia Pacific look to simplify and earn rewards through their daily purchases to spend on travel through a single loyalty scheme.

The pre-pandemic road warrior was armed with such loyalty programmes and co-branded credits to earn as many points as possible with many business travellers playing the system. Yet the frequent flier programmes (FFPs) are in full target of no-fly campaigners. The call for a ban on FFPs' miles and points is falling on deaf ears, but as net zero targets come into focus by 2030 and beyond, these programmes may find themselves under legislative scrutiny, requiring a major change in approach to what and how to reward consumers.

Next generation loyalty is already making its presence felt such as NFTs where rewards are digitalized as tokens, and benefits can be redeemed virtually or in the real world to drive brand engagement.

TIPS

- When it comes to consumers, ease, flexibility and convenience resonate among the most desirable features to help achieve their travel goals.
- With the complexity of the travel ecosystem, payments are the glue that binds the different handovers together, for an integrated lifestyle and consumer-centric experience.
- Global brands are targeting payments as a means to find value creation for themselves and their consumers.
- Finance and payments are constantly being disrupted by new technologies like AI – keeping ahead of future innovations and next gen finance models are key.
- There are multiple ways to deliver additional layers of personalization thanks to payments and loyalty intersecting with climate impact.
- Increasingly the sustainable success of travel businesses will take place at the nexus between climate action and digitalization.

Travel technology stacks and the competitive B2B ecosystem

Travel technology is a fragmented sector with hundreds of thousands of businesses providing business to business (B2B) services, powering brand operations and managing customer relationships. Legacy systems such as the GDS continue to evolve but also hinder evolution to meet the needs of e-commerce-experienced consumers. Global GDS players, Amadeus, Sabre and Travelport continue to evolve in line with the needs for their travel business clients. Behind the scenes, running a travel business is a highly complex undertaking, requiring a joined - up approach across multiple systems.

In aviation, new distribution capability (NDC) has been developed to bypass legacy systems and transform airlines into retailers, enabling the sale of services beyond flights to deliver a more personalized offer to consumers including ancillaries.

Hotels invest millions of dollars in customer relationship management (CRM) systems, revenue management solutions (RMS), property management systems (PMS) and central reservations systems (CSR). These solutions have all moved or are in the process of moving into the cloud, delivered as software as a service (SaaS). Not surprisingly, the big GDS players are investing in and acquiring B2B travel tech players, such as Sabre's acquisition of Genares in 2024 as they transform their business models. In aviation and hospitality, AI-first players like Flyr that use deep learning are wiping the slate clean and breaking the link with legacy systems to drive revenue optimization.

The role of RMS is to optimize pricing to deliver the strongest revenue per available room (RevPAR) growth, a complex balancing act of percentage occupancy and average daily room rate (ADR). It is based on yield management, leveraging supply and demand analytics.

Leading hotel tech players include SiteMinder, Canary, Cloudbeds, Lighthouse and Mews.[43] SiteMinder has the largest hotel distribution ecosystem, working with players like Trip.com, Airbnb, Expedia, Booking.com, Google HotelAds and GDS working with over 41,00 hotels like Hyatt and Best Western.

REAL-WORLD EXAMPLE

Mews – hospitality cloud delivering automation for an elevated hotel experience

Mews is a cloud-native B2B software as a service (SaaS) provider providing property management system (PMS) solutions to hospitality businesses. Its thousands of clients include some of the world's leading hotel companies like Strawberry Group (formerly Nordic Choice) as well as hostel and apartment brands like Generator-Freehand.

The Czech-founded company was born out of a need that was unmet for providing the necessary integration between PMS and related service partners that incumbent PMS players like Oracle would only provide for a costly fee. Mews' cloud-native and open API approach ensures that it can adapt and move in a responsive and agile way to technological advances like Gen AI and in response to changing consumer trends.

Dr Wouter Geerts, Director of Market Research and Intelligence at Mews, explains that the travel tech stack that many hotel businesses rely on is antiquated, based on legacy systems. Geerts continues that there are a lot of moving parts in hospitality meaning that tech players are playing catch up to move their solutions into the cloud.

Mews is cloud-native and its marketplace provides integration with over 1,000 leading tech partners on its PMS platform. With this open approach, Mews' hospitality customers can build their own tech stack to suit their needs, from integrating channel management to revenue management and sustainability tools, all on top of Mews' cloud platform. One of Mews' integrations is Hotels for Trees where every time a guest opts out of room cleaning, a tree is planted. Consumer-centric services include virtual concierge, digital key for contactless check-in, booking engine and seamless payments.

Mews aims to deliver 'heads-up hospitality' encouraging staff to spend less time on screens and more time providing guest experiences. Half of the 1,000 employees are focused on R&D and product development with foresight on emerging consumer trends through the anonymized data gathered on guests' profiles.

'AI is definitely going to make a big impact for guest personalization, whilst simplifying and automating operations for hotel staff. We are just at the beginning with a huge amount of potential for this technology, removing menial tasks and replacing them with elevated service interactions between staff and guests for a quality human touch.'

Dr Wouter Geerts, Director of Market Research and Intelligence at Mews

Hospitality at its core is about the guest experience where the best experiences are created through human interaction, where technology like Gen AI can improve efficiency and service delivery. AI entails knowing the customer, their preferences and behaviours ahead of time, ensuring a personalized and seamless experience that is relevant and contextualized across the customer journey. Players like Mews are delivering the digital know-how to leave hoteliers to do what they do best – delight and surprise their guests.

Decentralized travel business models

Decentralized travel companies are emerging, cutting out the middle man and breaking out of the traditional distribution model leading to the next stage of disintermediation. Decentralized autonomous organizations (DAOs) and decentralized autonomous corporations (DACs) reject centralized control typical of most businesses, instead working bottom up with members having voting rights, united by a common purpose. Decentralization is enabled by blockchain such as Ethereum for transparency and greater security through smart contracts.

In the same way that OTAs disintermediated travel agencies, and Airbnb disintermediated OTAs, a new phase of disruption is on the cards with a shift away from centralized control. The opportunity for peer-to-peer (P2P) networks is huge in travel where travel suppliers, communities and member groups can take charge of their own travel products and services. Equally this could lead to the next generation of social commerce, democratizing travel booking through individuals such as influencers. This marks a paradigm shift in how travel businesses could potentially operate and engage with consumers where every party will own their own digital data and third party fees will be relegated to the past.

Expedia launched Travel Shops for creators and influencers as a first-of-its-kind platform connecting influencers and travellers for tips and travel booking through referral links, whereby creators earn commission. Each marketplace is unique and customizable leading to a new era of where anyone can promote and sell travel.

Transforming to an open, permissionless and secure online travel marketplace could empower greater levels of choice, authenticity and personalization for travellers guaranteed by smart contracts. Democratizing the sale of travel products and services opens up new opportunities.

Travala.com is one of the first next-generation crypto-native, web 3.0 travel platforms powered by blockchain, selling flights, hotels and experiences accepting over 100 cryptocurrencies, with the aim to democratize travel. Its Smart Diamond loyalty scheme includes NFT rewards along with real-world benefits like access to lounges, concierge services and earn points with Marriot Bonvoy. Next gen features include NFTs as rewards to mark proof of travel, and various drops for digital or real-world experiences.

Anyone can become a travel supplier or OTA through web 3.0 and blockchain solutions like Xeni. Xeni partners with social influencers as well as global OTAs like Expedia, Rakuten and Webbeds providing access to wholesale travel prices for resale.

Dtravel is a decentralized travel company specializing in P2P short-term rentals in order to drive direct bookings. On its marketplace it launched rental night bookings as NFTs, tokenizing bookings via smart contracts. Tokenization of real-world assets as Nite Tokens is possible through partnership with Polygon. The benefit to travellers is that they can exchange or resell their token for greater flexibility, while the rental owner receives commission from this new revenue stream of secondary sales. The company claims to launch the first global open-source booking protocol for short-term rentals, removing layers of complexity through multiple APIs. Dtravel is at the intersection of travel and blockchain, and its TRVL cryptocurrency can be collected and even exchanged for physical fiat currency by its members including companies and individuals.

In future, if every decentralized marketplace launches its own digital currency then this would result in a highly fragmented payments landscape. Decentralizing contracts will also open up new opportunities for the resale of travel products and services. Cryptocurrency is already used for purchasing travel, with online booking sites like Alternative Airlines accepting over 40 different types of payments options including over 70 different cryptocurrencies, BNPL and e-wallets for payments.

Web 3.0 blockchain players like Chain4Travel are making inroads into travel with its Camino Network, partnering with players like Lufthansa and Der Touristik Online to run hackathons to innovate and find ways to scale adoption of areas such as SAF on its blockchain ecosystem.

There is massive potential for a new travel unicorn to emerge from web 3.0 start-ups working in blockchain and decentralized commerce. Pushing out the middleman is an ever-green goal for travel where direct sales are increasingly achievable thanks to decentralized players. The question is, who will be the new leaders of the new platforms?

TIPS

- Decentralization thanks to technology like blockchain is nascent yet could unlock much potential for travel businesses and consumers by bypassing the middlemen and their fees.

- De-finance and crypto suffer from a bad name rightly due to fraud and scams, so moves into this new space should be taken with great caution, working with established players.

- The shift to decentralized platforms is when there is distrust in the current systems, where a more human-centric and trust-based approach is required.

Summary

Travel by nature is in a constant stage of digital transformation and as a result, is in flux. Technology is fundamental to all major back-end operations, service delivery and ultimately more woven into the travel experience. If the technology fails at any point in the chain, then it risks impacting the joy of experiencing new cultures, places and communities leading to a bad experience for the traveller. There is a lot riding on the technology working to keep the travel ecosystem functioning without travel chaos, disruption and delays or millions of missed revenue opportunities due to patchy Wi-Fi. Due to the fragmented nature of the industry, many travel businesses continue to work with legacy systems in the back-end.

Technology powers innovation across the customer journey, opening up new ways to engage with consumers, drive greater levels of personalization and tap into new revenue streams. However, digitalization is not without risks where data privacy and security are fundamental pillars that must be adhered to.

Not all technologies will succeed. With developments in web 3.0 and consumer-centric privacy regulations, digital is no longer the wild west as regulators play catch up and consumers are empowered to take back control of their digital footprint and relationships with brands. In line with growing decentralization, consumers are even more in the driving seat in terms of with whom and how they engage with businesses. This opens up opportunities to create more meaningful and emotional connections with consumers based on shared values, where the balance of power in the customer-brand relationship is more equal. Consumers decide what trade-offs they are willing to make for receiving greater levels of service and personalized travel experiences. Time-saving and convenience continue to be key attributes for consumers to ensure that the hassle factor of travel is minimized.

Yet to truly deliver amazing travel experiences, technology needs to be invisible so that the joy of immersion in the real world, whether engaging with people, nature or heritage and culture, is maximized.

The ongoing digital transformation that is required to ensure safe, seamless and secure travel equally plays a pivotal role in delivering operational efficiencies to meet the necessary net zero targets. Technology such as Gen AI, advanced analytics and automation help to improve operations through constant evaluation and innovation. Digital technology and data have the power to encourage consumers to make the best choice for the planet when selecting their travel providers, provide smart mobility options and greater transparency of carbon emissions and impacts.

Digital technology is earmarked as a key driver of SDGs helping to reduce delivering the costs of delivering the SDGs by up to US\$56 trillion by 2030, enabling 103 out of 169 targets.[44] Deployed responsibly, technology can lift people out of poverty and empower them to achieve their full potential. Failure to embrace digital transformation for the greater good is tantamount to failure in protecting the planet for future generations.

There is no silver bullet, however, where the carbon emissions from digital activities are estimated to account for the same as the aviation sector at around 2.3 per cent to 3.7 per cent.[45] Bitcoin, the most famous cryptocurrency, produces more carbon emissions than some countries through energy consumption while mining.[46] Gen AI and next generation web 3.0 such as blockchain, cryptocurrency, NFTs and metaverse are highly carbon intensive. Taking a holistic view of all negative climate impacts including digital is necessary.

Gen AI has unleashed new challenges in terms of perpetuating bias and misinformation so new regulations, checks and balances need to be enforced. Future digitalization across the customer journey also requires travel businesses to tread a fine line with privacy laws and changing consumer attitudes while delivering on sustainability targets. Digital innovation can elevate the sheer joy of nature, cultural and climate positive experiences if tech is used for good.

Endnotes

1 Expedia Group (2023) partner.expediagroup.com/content/dam/unified/partner/documents/reports/2023-reports/report-path-to-Purchase-2023-final_en-us.pdf (archived at https://perma.cc/2PKU-FLYA)

2 Euromonitor International (2024) www.euromonitor.com/ (archived at https://perma.cc/8CH2-JQ3B)

3 Ericsson (nd) https://www.ericsson.com/en/reports-and-papers/white-papers/a-research-outlook-towards-6g (archived at https://perma.cc/693F-5SF4)

4 World Bank from ITU (2024) data.worldbank.org/indicator/IT.CEL.SETS.P2 (archived at https://perma.cc/G4T4-H7AD)

5 Phocuswire (2023) www.phocuswire.com/influencer-marketing-windstar-return-on-investment-jerne (archived at https://perma.cc/PD3G-A2BH)

6 Smart Insights (2024) www.smartinsights.com/search-engine-marketing/search-engine-statistics/ (archived at https://perma.cc/69Z5-FSLP)

7 Phocuswire (2024) www.phocuswire.com/online-travel-giants-new-record-marketing-spend-in-2023 (archived at https://perma.cc/7NVY-GGMS)

8 Google (2024) blog.google/products/search/google-summer-travel-tips-2024/ (archived at https://perma.cc/55AS-4AH2)

9 AI Business (2023) https://aibusiness.com/nlp/chatgpt-passes-1b-page-views (archived at https://perma.cc/GM2U-CANR)

10 Lufthansa Innovation Hub (2023) tnmt.com/newsletter-snippets/ai-investment-trends-in-travel-and-mobility-tech/ (archived at https://perma.cc/TD2B-QCHW)

11 Travel Weekly (2023) www.travelweekly.com/Travel-News/Travel-Technology/ Tripadvisor-sees-dollar-signs-generative-AI (archived at https://perma.cc/9LXL-V5NP)

12 Web in Travel (2024) www.webintravel.com/with-innovationism-as-its-new-philosophy-trip-com-group-comes-roaring-back/ (archived at https://perma.cc/EZ26-Y4QX)

13 Web in Travel (2024) www.webintravel.com/with-innovationism-as-its-new-philosophy-trip-com-group-comes-roaring-back/ (archived at https://perma.cc/8BUH-88X6)

14 Amadeus (2024) amadeus.com/en/blog/articles/how-travel-sellers-can-use-technology-to-achieve-a-more-sustainable-future (archived at https://perma.cc/MVQ3-ZXGV)

15 Fortune (2023) fortune.com/2023/06/08/sam-altman-openai-chatgpt-worries-15-quotes/ (archived at https://perma.cc/Y53V-QZET); Fortune (2024) fortune.com/2024/02/13/sam-altman-openai-ai-subtle-societal-misalignments-killer-robots (archived at https://perma.cc/89CH-U35W)

16 Pew Research Center (2023) www.pewresearch.org/short-reads/2023/12/08/ striking-findings-from-2023/ (archived at https://perma.cc/48DF-VC34)

17 Capgemini (2023) www.capgemini.com/news/press-releases/73-of-consumers-globally-say-they-trust-content-created-by-generative-ai/ (archived at https://perma.cc/PKC7-7N5Q)

18 Amadeus (2024) amadeus.com/en/insights/press-release/travel-tech-investment-trends-2024 (archived at https://perma.cc/5NRN-VW6T)

19 Ibid

20 Goldman Sachs (2023) www.goldmansachs.com/intelligence/pages/generative-ai-could-raise-global-gdp-by-7-percent.html (archived at https://perma.cc/C5FH-XMWG)

21 TechTarget (2024) www.techtarget.com/searcherp/feature/AR-vs-VR-vs-MR-Differences-similarities-and-manufacturing-uses (archived at https://perma.cc/5BDW-6WSF)

22 Snapchat (2023) forbusiness.snapchat.com/blog/uncovering-the-value-of-augmented-reality-in-travel (archived at https://perma.cc/G6VV-TMHR)

23 Gartner (2022) www.gartner.co.uk/en/articles/what-is-a-metaverse#:~:text=The%20 Metaverse%20is%20a%20collective,enhanced%20physical%20and%20 digital%20reality (archived at https://perma.cc/MU6E-J772)

24 Unity (2023) blog.unity.com/industry/how-digital-twins-are-transforming-large-scale-airports (archived at https://perma.cc/F295-3FVH)

25 Eurocities (2024) https://eurocities.eu/stories/barcelona-shapes-the-future-of-city-planning/ (archived at https://perma.cc/TF5Q-ABM2)

26 M Wandres, A Espejo, T Sovea, S Tetoa, F Malologa, A Webb, et al. (2024) A national-scale coastal flood hazard assessment for the atoll nation of Tuvalu. *Earth's Future*, 12 (2024) doi.org/10.1029/2023EF003924 (archived at https://perma.cc/777P-HCU7)

27 Airbnb (2024) investors.airbnb.com/financials/sec-filings/sec-filings-details/ default.aspx?FilingId=17283799 (archived at https://perma.cc/ZZ6R-AWZ7)

28 IATA (2023) www.iata.org/en/iata-repository/pressroom/presentations/ global-passenger-survey-gmd2023/ (archived at https://perma.cc/7G6U-UY99)

29 Ibid

30 IATA (2023) www.iata.org/en/iata-repository/pressroom/presentations/ global-passenger-survey-gmd2023/ (archived at https://perma.cc/7G6U-UY99)

31 Skytrax (2024) www.worldairportawards.com/worlds-best-airport-security-2024/ (archived at https://perma.cc/3ALM-KR7C)

32 Skytrax (2024) www.worldairportawards.com/worlds-best-airport-immigration-2024/ (archived at https://perma.cc/TA6S-YBQ9)

33 US Travel (2024) www.ustravel.org/press/stunning-new-research-ranks-united-states-nearly-dead-last-competition-global-travelers (archived at https://perma.cc/TW7G-WV54)

34 SITA (2023) www.sita.aero/globalassets/docs/surveys--reports/2023-air-transport-it-insights-v6.1.pdf (archived at https://perma.cc/2C7A-JL8T)

35 Ibid

36 Hospitalitytech (2023) hospitalitytech.com/research-finds-54-hospitality-organizations-have-increased-cybersecurity-budgets-last-12-months (archived at https://perma.cc/UC7M-E7TC)

37 Visa (2023) investor.visa.com/news/news-details/2023/Revenge-Travel-is-Here-to-Stay-Visa-Study-Reveals/default.aspx (archived at https://perma.cc/4YZD-WWWQ)

38 Visa (2023) www.visa.com.sg/about-visa/blog/bdp/2023/11/27/contactless-payments-experiential-1701068432525.html (archived at https://perma.cc/XJ84-2KU2)

39 Atlantic Council (2024) www.atlanticcouncil.org/cbdctracker/ (archived at https://perma.cc/E2CA-CKNP)

40 Business of Apps (2024) www.businessofapps.com/data/wechat-statistics/ (archived at https://perma.cc/2F8P-9GXT)

41 Visa (2022) usa.visa.com/visa-everywhere/blog/bdp/2022/04/20/more-sustainable-future-1650496057162.html (archived at https://perma.cc/3L7S-TXBK)

42 Marriott (2024) news.marriott.com/news/2024/03/21/homes-villas-by-marriott-bonvoy-makes-finding-a-vacation-home-easy-with-gen-ai-and-intuitive-with-innovative-search-tool-powered-by-generative-ai-launching-today (archived at https://perma.cc/XMR6-BS6A)

43 Hotel Technology Newsc (2024) hoteltechnologynews.com/2024/03/hoteltechreport-announces-its-picks-for-the-best-hotel-tech-apps-of-2024/ (archived at https://perma.cc/QTD8-7V9Q)

44 International Institute for Sustainable Development (2023) sdg.iisd.org/news/force-for-good-report-explores-role-of-technology-in-helping-reach-sdgs/ (archived at https://perma.cc/YB6Z-FAX6)

45 MyClimate.org (2024) www.myclimate.org/en/information/faq/faq-detail/what-is-a-digital-carbon-footprint/ (archived at https://perma.cc/JQ3M-P3GS)

46 Plan A (2024) plana.earth/academy/how-can-web-3-0-cryptocurrency-be-solution-climate-change (archived at https://perma.cc/6LF6-JBWK)

05

Sustainable technology solutions for the next generation of travel

Introduction

With increased urbanization, digital transformation through advances in Gen AI and supercomputing, a new era of technology will emerge and impact travel and tourism. Converging with climate and biodiversity challenges, new technology is creating ever more opportunities to fix the challenges that humanity faces.

Against the backdrop of a planetary existential crisis – even if global temperatures are kept at 2°C warming – untold costs and damages will be inflicted on lives and economies. Two billion people could be at risk of extreme and dangerous heat (over 29°C) by 2100 if climate targets are not met where the world is on track to reach 2.7°C global warming compared to pre-industrial levels.[1] The poorest will suffer the heaviest burden. There is a moral and ethical obligation for government policies, businesses and technology to be deployed to ensure a sustainable and fair transition for all.

Transport, energy and food are targeted for areas for accelerating positive tipping points to counter negative climate tipping points. Travel and tourism are heavily reliant on transport and infrastructure that are major carbon emitters and the focus of many sustainable innovations to deliver net zero solutions.

The siloes between sustainability and digital transformation will continue to disappear, forced to merge under ever-growing pressure to tackle the climate emergency. Interdisciplinary collaboration at scale is required. Digital and green innovation are critical for the transition to a net zero world, but equally entail negative impacts in terms of energy use if not powered by renewables. Next generation technology will need to integrate sustainable and climate-ready features to the foundations of the built environment, products, services and experiences.

The new era of hyperconnectivity powered by the next generation of internet – 6G – will lead to super-fast speeds of computational power. Although it is hard to predict what new types of technology will emerge, travel businesses will benefit from higher levels of operational efficiencies. Ever more data on ESG impacts will be recorded, while consumers will enjoy ever greater personalization and contextual information with services at the right time and place across the customer journey.

In the way that mobile connectivity has spread with near ubiquity, Gen AI, robotics and automation, edge computing and IoT will be pervasive. Future generational cohorts and businesses will not just be tech-savvy; they will be AI native. With everything, not just mobility, provided 'as a service' including travel itself, a new era of quantification is coming, with higher levels of transparency and accountability. Yet trust, quality, purpose and safety must be preserved.

Futuristic transport solutions such as the hyperloop and eVTOLs – so-called air taxis – will become standard worldwide and play a key part of the multi-modal transport system. Solving the last-mile challenge will be in sharp focus as seen in retail. Decarbonization of fleet from electric vehicles and ships to aircraft will be accelerated, while suborbital space travel will open up long-haul destinations like Australasia and Asia Pacific.

Scope 3 emissions are the hardest to abate, leading to the need for stronger partnerships outside of the travel industry, especially energy and construction as a full lifecycle approach is increasingly adopted to the built environment. Taking a holistic system-wide approach that is transparent about the trade-offs required will help to navigate the uncertainties of an ever-faster digitalized world that faces ever greater dangers as new tipping points are breached. Travel and tourism are at the heart of the problem, and equally part of the solution. Young people especially understand the benefits that travel can bring; the author's 12-year-old daughter, Ingrid, shares her view in Figure 5.1.

FIGURE 5.1 Message from Ingrid

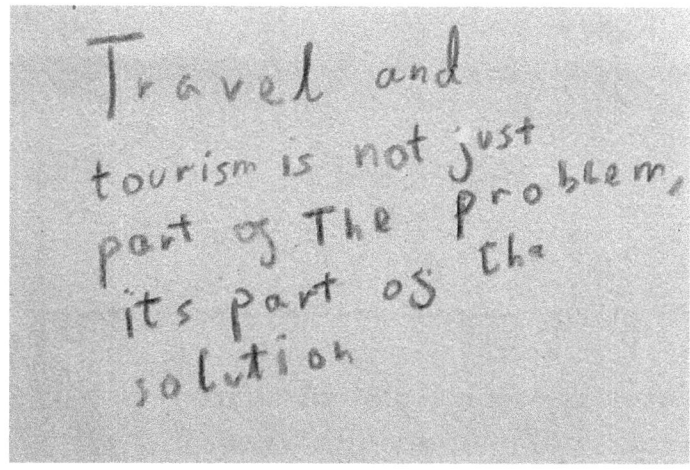

Hyper-connected world

New era of hyperconnectivity

The pace of technological change is gaining in speed, where leaps forward create opportunities and challenges for humankind. With AI entering an important new phase of development with Generative AI, it will help accelerate innovation by combining with humans to find solutions. While Kurzweil's Law of Accelerating Returns points to the non-linear development of technological transformations and that changes will happen exponentially.[2] These future paradigm shifts are hard to predict, but already technology giants and their innovation and research and development (R&D) work give clues to what may lie ahead, where futurists like Kurzweil also point to the singularity where AI exceeds mankind's combined cognitive powers with Artificial General Intelligence (AGI) leading to the need to adopt responsible AI business practices and codes of conduct.

The next generation of wireless communications, moving from 5G to 6G capabilities, will exponentially boost communications between people, devices, machines and sensors, collecting unimaginable reams of data. This will lead to new ways of working, communicating and living. 5G mobile technologies already support smart cities, smart homes, the cloud, AR/VR, machine-to-machine communications and autonomous transport solutions. 5G is still not fully rolled out so there is still some way to go before 6G becomes a reality. 6G is expected to roll out around 2030, with Asia Pacific expected to be first.[3] The leap from 4G to 5G was exponential in terms of connectivity, with the leap to 6G taking internet speed to a whole new level, over 100 times faster.[4]

With 6G's omnipresent connectivity and unprecedented faster speeds of communication – not just between humans but through edge computing and devices – there will be a digital–physical continuum. Projects like Hexa-X II in the EU will ensure that the new capabilities provided by 6G will be used to ensure a sustainable, inclusive and trustworthy 6G ecosystem.

The race to 6G is heating up between China and the West, the US signing a pact with its allies including Canada, Australia, Czech Republic, Finland, France, Japan, South Korea and the UK to establish 6G principles based on trust and security. Key to next gen technology like 6G is energy efficiency, affordability and digital transformation to drive sustainability across industries.

China is treading its own path. The country is investing heavily in R&D in 6G and launched a low earth orbit (LEO) satellite to test ground to space communication, that will combine with 6G networks. This combined approach will deliver ubiquitous connectivity through a universal integrated 'Internet of Everything' on land, air, sea and space. This Internet of Everything goes beyond the Internet of Things to connect people to machines, machines to machines and people to people with four pillars: people, processes, data and things.

To deliver breathtaking speeds, internet speed per second will move from gigabytes to terabytes, i.e. billion to trillion, greater bandwidths will be delivered, requiring new 6G infrastructure to power them.

6G aims to deliver greater bandwidth and lower latency ensuring that connectivity is truly universal and persistent, operating at higher frequency bands. This opens up multiple new opportunities for travel businesses and destinations to track, monitor and improve their operations and sustainability performance, while finding new unique and immersive experiences for their customers.

Yet advances do not stop with 6G. Quantum computing and a quantum internet are being tested in areas such as smart cities, with trials in the US, China and the Netherlands. Quantum computing is not intended to replace traditional networks, instead working alongside and providing higher levels of capabilities for complex problems such as making cities sustainable.

TIPS

- As with all iterations of wireless technology, the step up to the next level of connectivity will not run smoothly; ensuring existing connectivity works is key with one eye on future developments.

- With new tech advancements, new cyber threats emerge requiring the next generation of cyber-security and advanced privacy standards.

- With new technologies taking many businesses by surprise, such as ChatGPT, R&D must not take a back seat in times of economic uncertainty.

- Travel businesses working in partnership with innovation clusters can open up new opportunities for shared benefits.

Future of transport and mobility

Integrated multi-modality

One of the main beneficiaries of the move to 6G is transport and traffic management in smart cities where 6G will be AI native, leading to advancements in digital twins and powering autonomous vehicles (AVs).

Worldwide, transport is one of the biggest emitters, globally accounting for 23 per cent of direct GHG, especially road transport accounting for 70 per cent of transport emissions as heavily reliant on fossil fuels.[5] Fundamental to achieving a climate-positive future is decarbonizing transport and moving further towards shared mobility solutions and promoting walking and cycling. A complete systemic transformation is called for to reach climate ambitions by embracing the circular economy, shared mobility, digital transformation, changing consumer behaviour and reducing transport use and boosting energy-efficiencies. Travel businesses and destinations are at the forefront of this transformation.

However, electrification of transport requires electricity from the grid network to come from renewable energy sources not fossil fuels in order that carbon emissions are not increased. Fuel mix is therefore critical to consider, with electrification going hand in hand with the shift to renewables.

Mobility as a service (MaaS) has received much attention for well over a decade from smart cities, investors and start-ups looking to solve the last-mile challenge and deliver a seamless, end-to-end multi-modal service on-demand. MaaS offers integrated solutions on a common platform for ease of booking, convenience and payment. Combined with electrification, the solutions are even more appealing and contribute to climate targets and smart city goals. The future for MaaS is ever greater interoperability so that there is openness and shared data amongst providers, public and private sectors through initiatives such as the Open MaaS Ecosystem. The ultimate goal is to provide a better service than consumers who own a car would receive.

Uber as the most famous MaaS player has spawned thousands of copy-cats looking to provide ride-hail, ride-share, bike-sharing and e-scooter hire around the world. The company continues to grow healthily, reporting 9.5 billion trips in 2023, worth US$138 billion, and delivering revenues of US$37 billion.[6]

Next generation of high-speed rail

Rail is one of the least polluting modes of transport and continues to attract attention in how to build a sustainable alternative to air travel to achieve climate neutrality. Leading high-speed rail are countries like France, Spain and Germany, Japan and China where the latter enjoys the fastest speeds in the world of 460 km/h with the Shanghai Maglev train. The future for high-speed rail looks bright, except in the UK with its infamous HS2 project.

REAL-WORLD EXAMPLE
India's ambitious future-ready plans for high-speed rail

In India, work is well underway on the Mumbai-Ahmedabad high-speed railway (HSR) that will be the country's first with a speed of 320km/h conducted by the National High Speed Rail Corporation Ltd (NHSRCL) over 508 km. The plan will ramp up the frequency of the bullet trains from 35 per day in the first year of operations to 105 three decades later.[7] NHSRCL is partnering with Japan's iconic Shinkansen technology system via the Japan International Cooperation Agency, as the HSR forerunner, and France's SNCF.

The investment is part of India's economic, environmental and social development as outlined by the National Rail Plan (NRP) 2030. Indian Railways has a long heritage since the first rail journey in 1853 and ongoing investment is part of the government's Make in India initiative to achieve its sustainable development goals and NDCs.

Reduced times for HSR travel will benefit consumers and freight, and reduce congestion on roads. There is a broader National Railway Plan for seven new corridors to feed into the HSR network. The adoption of HSR is a key lever in reducing emissions especially in the world's most populous nation, where even inter-city rail is 3.5 times less polluting than air travel per passenger.[8] As part of the HSR infrastructure, a multi-modal transport hub is planned for Sabarmati for promoting seamless travel and connections. As with all infrastructure projects, there are delays in completion with 2026 touted as the revised launch with costs sure to rise.

Bullet trains are just one of the key pillars of the NRP, with Prime Minister Modi announcing the production of 4,500 new Vande Bharat trains by 2047, that are long-distance with potential speeds of 250 km/h. Forecasts for rail passengers are expected to rise from 9.5 billion in 2021 to 15.5 billion in 2041 and almost hit 20 billion by 2051.[9] The majority of locomotives will be electric trains, but a small percentage will continue to be diesel, which seems to run contrary to the overarching need to decarbonize. However, electrification is taking place at a strong pace across the network. India is on track to achieve rail net zero emissions by 2030 with 94 per cent electrified already, compared with Amtrak in the US that has only 1 per cent of its rail network electrified and the EU at 56 per cent.[10] Nevertheless, the source of electricity from the grid equally needs to transition to renewables.

With rail as a catalyst for decarbonizing transport, large cross-border projects such as the GCC Railway are back on track despite a rocky start, planning to link all six GCC countries. One of the largest investment projects has not been easy to keep to target, but agreement is for the 2,000+ km rail network to be open by 2030 at a cost of US$250 billion.[11]

Yet the ambition does not stop there. At the G20 in India, a new India–Middle East–Europe Corridor was agreed upon, providing cross-regional connectivity, leveraging rail and sea transport. This initiative is to compete directly with China's Belt and Road Initiative.

The Africa Union's 2063 Agenda includes the flagship African Integrated Railway Network – a critical enabler of the region's sustainable, inclusive development and African Continent Free Trade Area (AfCFT). Progress is slow, but Morocco was the first on the continent to enjoy high-speed rail in 2018. Launched the previous year, the Chinese built the Addis Ababa (Ethiopia)–Djibouti Railway, part of the Belt and Road Initiative.

> **TIPS**
>
> - Existing infrastructure can be transformed through electrification of rail and other transport modes; however, it is also essential to transform the national grid by transitioning to renewables.
> - The shift to rail must be accompanied by a consumer-centric, seamless passenger experience.

Beyond high-speed rail – Hyperloop receives a shot in the arm

Over the past decade, hyperloop has enjoyed much hype yet has struggled to bring Elon Musk's vision to life, delivering the next generation of transport at twice the speed of high-speed rail. Despite the high-profile failure of Hyperloop One that closed in 2023, which had DP World and Virgin as investors, there remains some interest in testing the technology. Using magnetic levitation (maglev), the system works with pods travelling at speeds of up to 700 mph/1,200 km/h operated in a vacuum tube thanks to the removal of air resistance.

China aims to launch a hyperloop between Shanghai and Hangzhou by 2035, for example.

Meanwhile, Hyperloop Transportation Technologies (HyperloopTT) aims to launch a Hyperloop in Italy, connecting Venice-Mestre and Padua. The Hyper Transfer project brings together partners such as WeBuild, RINA and Leonardo for the Venice commercial prototype. Within the capsule, there is capacity for 30 passengers, focusing on a class-free, human-centric experience powered by digital and biometrics. Services on offer include luggage space, entertainment, food and washrooms. The Italy prototype involves public and private partnership, with an investment of US$868 million, with the goal to be in operation by 2030.[12]

Despite the technology being 10 years old and not yet commercially viable, the EU is banking on hyperloop to help transform to sustainable mobility with the opening of the European Hyperloop Centre and the Hyperloop Development Fund. With journeys dramatically cut, if start-ups like HyperloopTT, Hardt (Netherlands), Zeleros (Spain) or Nevomo (Poland) can make it commercially, it would mark a true revolution for the future of transport.

Despite the setbacks with Hyperloop One where Dubai Ports was an investor, the UAE and Saudi Arabia see a future for hyperloop as part of their visions for the future of integrated multi-modal transport across the region. The HyperloopTT Italy project has sparked fresh hope for the technology to succeed.

TIPS

- Just because the technology can be built does not mean it should if it is not inclusive in terms of ticket price for consumers, cost-effective to build or causes negative impacts.

- Scope 3 emissions need to be considered for any infrastructure projects.

- Failure is part and parcel of innovation, so a pragmatic mindset and iterative approach are best.

- Some technologies arrive before their time but may reappear at a later date.

Urban air mobility ready for take-off

With aviation being one of the hardest sectors to abate for carbon emissions, and the long journey to achieve net zero by 2050, the urban air mobility (UAM) sector is garnering much attention from governments, investors, urban planners and smart cities. In a positive forecast scenario, the market for advanced air mobility (AAM) could reach EUR55 billion.[13] With the switch to electric, there are reductions gained from electrical vertical take-off and landing vehicles (eVTOLs) in terms of carbon emissions compared to internal combustion engine vehicles (ICEV), namely 35 per cent lower.[14] This includes developments of drones as well as electric air taxis expected to fly in low-level airspace at 500–1,000 feet.

In the US, the Federal Aviation Administration (FAA) and National Aeronautics and Space Administration (NASA) are working with UAM companies to develop the sector in a safe way, where new locations not previously served by air travel will benefit from AAM solutions. Regional air mobility is a new core focus for the FAA. This will help to solve the last-mile conundrum and ensure that there is point-to-point intermodal transport options that are climate friendly.

There are many companies investing and developing eVTOLs. Already there have been some high-profile exits such as Airbus closing down Voom, and Uber wrapping up its Uber Elevate programme, instead investing in Joby. Joby is the world's first UAM unicorn after receiving investment from Toyota in 2020. Once it receives approval from the FAA its ridesharing air services will be bookable on its own app, via Uber and even via Delta Airlines for airport shuttles.

Ahead of the pack in terms of design and approval is Volocopter (Germany), the first to receive certification from the European Union Aviation Safety Agency (EASA). It is moving from design and testing, and once it receives certification will move into mass manufacturing. Its VoloCity vehicle design is for short intra-city trips of 35 km for two passengers, with a speed of 110 km/h.[15] Singapore, Helsinki (Finland), Hamburg (Germany), Paris (France) and Dubai (UAE) have already held test flights with Volocopter's air taxi, VoloCity, with much interest in the benefits in terms of the digital and green transition required. Expansion is global. Volocopter aims to expand in Japan and NEOM (Saudi Arabia).

In Taiyuan (China), the city is investing in a fleet of 500 pilotless eVTOLs manufactured by EHang with a planned use for tourism and transportation. The company aims for a climate-friendly integrated transport system, that runs point to point, reduces congestion on the roads and delivers a seamless travel experience end to end for consumers.

Lilium of Germany is listed on Nasdaq and focuses on regional eVTOL flights, with strong partnerships including Netjets, AZUL, Saudia, ABB and Volare, with global interest in its aircraft. The company is gearing up for launch in 2026 when it hopes to receive dual certification from FAA and EASA. Singapore launched the first vertiport prototype in 2019 at Marina Bay, in partnership with Skyports and Volocopter.

The shift to eVTOLs is complex due to ensuring safety and not causing negative impacts on local communities through increased air traffic and noise. The eVTOL ecosystem of maintenance, support and infrastructure needs to be built for the network of vertiports.

eVTOLs are expected to be as familiar a sight in the future multi-modal transport system whereas for the masses they remain a futuristic concept, but still not as alien as flying cars. UAM is expected to be driven by increased urbanization, resultant congestion, the rise of autonomous electric vehicles and aircraft as catalysts of the net zero transition. The first stage of UAM will focus primarily on intra-city and intra-metropolitan areas, with longer range distances restricted by battery capacity.

As battery power increases, this will open up short-haul city-to-city transit, providing relief and saving time for commuters. However, electric batteries will also restrict the distance and range of eVTOLs due to the need to recharge.

The competition from incumbent manufacturers is heating up. Airbus despite its failure with Voom continues to push ahead in the UAM space, debuting its CityAirbus Next Gen aircraft with commercialization on the cards as it also met EASA approval.

UK start-up Lyte Aviation is taking a leap forward by aiming for a capacity of 40 passengers and four crew, designing a hybrid electric hydrogen-powered VTOL, going beyond intra-city to inter-city and regional flights with its SkyBus concept. It aims to cover distances up to 1,000 km with a speed of 300 km/h.[16]

REAL-WORLD EXAMPLE
Eve Air Mobility – designing the next generation of electric aviation
for urban mobility

Eve Holding Inc is one of the companies pushing the boundaries of aviation, through its strategic partnership with Brazilian aerospace manufacturer, Embratur. It is developing eVTOLs for launch in 2026, along with providing maintenance and developing a new Urban Air Traffic Management (UATM) system. It is testing its concept of operations in cities such as London (UK), Chicago and Miami (USA), Rio de Janeiro (Brazil) and Melbourne (Australia). This is vital to ensure that UAM low-flying aircraft can operate safely with the existing aviation system.

Eve plans to provide services and support, and not operate the aircraft instead leasing with already 2,850 vehicles on the books, worth US$8.6 billion.[17] Working with third parties, its clients include ridesharing platforms (Blade Urban Air Mobility, Flapper, Helipass), aircraft leasing companies as well as helicopter operators and airlines such as United Airlines. Partners include vertiports such as London City Airport, Heathrow Airport and Skyports.

The addressable market for Eve and its clients is the intra-city and intra-metropolitan areas, where its eVTOLs are designed to carry four passengers and one pilot, or six passengers in an autonomous aircraft.[18] The range is 100 km, thanks to its lift and cruise design.

The case for eVTOLs versus helicopters is compelling, where Eve states a 90 per cent noise reduction, and savings of 65 per cent for direct operating costs and 85 per cent saving for autonomous eVTOLs. This suggests that eVTOL travel will be more inclusive and less cost-prohibitive than standard aviation, which will increasingly face carbon taxes. Safety and certification will be paramount to ensure consumers are willing to upgrade from terrestrial to air commuting. The eVTOL market is highly competitive, combining specialist UAM developers (Archer, Volocopter, Joby) as well as aircraft and automotive manufacturers like Airbus, such as Honda and Hyundai.

Partnerships are critical to success. Eve is working with low-cost carrier (LCC) Jeju Air in Jeju island (South Korea) on developing UAM and accelerating commercial net zero air travel as part of its Carbon Free Island 2030 strategy. The pathway starts with sightseeing flights, progressing to air mobility and finally air shuttle services. Jeju already benefits from a strong electric charging network and electric buses and supporting electric grid network.

Consumer appetite in Jeju is not huge: key reasons for using UAM include 30.4 per cent price, 29.6 per cent convenience, ease 20 per cent and speed 13 per cent.[19] As Jeju Air and Eve move through the different phases, flights will transition from scheduled to on-demand UAM services matching supply and demand in real time. AI will be used to streamline and automate passenger and baggage screening.

In London, Eve is a member of the UAM consortium including Skyports, Volocopter, NATS and Heathrow and City airports. eVTOLs will link London Heathrow to London City Airport with vertiports co-located there, where the city will host four vertiports with 520 flights per day connecting financial hubs like Canary Wharf.[20] For example, Skyports' heliport at Canary Wharf will be part of the vertiport network and is already actively testing.

Operations are at risk of disruption from weather events, and with climate change effects increasingly extreme this may be more common. Consumers experiencing turbulence and motion sickness may be a downside that travellers are not willing to endure. By 2030, the value of UAM could reach US$10 billion, a fraction of the US$1 trillion forecast to be spent on shared mobility solutions.[21]

FIGURE 5.2 Eve Air Mobility

SOURCE Eve Air Mobility

TIPS

- With shared mobility solutions moving into the skies in future, coupled with the ever-increasing presence of delivery drones, the skies could become more crowded with potential for more accidents.

- Ensuring safe, sustainable and affordable flights will be key to uptake from consumers.

- There is an opportunity for travel and luxury brands to branch into urban mobility with their strong branding.

- With airlines' experience in safety, this will be a USP for airline-branded eVTOLs.

- Urban city planners as well as travel businesses will benefit from innovative, new partnerships with eVTOL players.

- The future of mobility is digital and electric.

Flying cars fast becoming a reality

Farfetched as it may sound, start-ups are racing to create the first flying car. Alef Aeronautics of the US backed by Elon Musk's SpaceX is leading the field. The concept was inspired by *Back to the Future* with the goal being to enable a vehicle that drives on roads and takes off into the air without a runway. The company even offers a 'pre-order a flying car' option on its website, with orders already over 2,800, worth US$300,000 per eVTOL able to transport two passengers.[22] Alef's Armada Model Zero flying car has been approved by the US FAA for special air worthiness with production scheduled in 2025. It will be powered by eight propellers, eight motors and batteries, and hydrogen can be included for additional cost.

However, the path to commercialization is fraught with challenges, and regulatory approval will be hard to achieve unless safety is guaranteed, receiving air and road worthiness approval. Another challenge will be the low speed at 40 km/h.

Competition is rife. Volkswagen Group China launched its flying car prototype, V.MO known as Flying Tiger in 2022. The goal is for autonomous flights, serving HNWI Chinese consumers with a capacity of four passengers and luggage, travelling 200 km/h for shuttle transfers.

KleinVision's AirCar hosted a famous first passenger – Jean-Michel Jarre – in April 2024 on its inaugural passenger flight with accompanying electronica soundtrack. Founder Stefan Klein's vision for AirCar started in 2017 in the Slovak Republic and received its air worthiness certificate in 2019. The company is licensing its technology to China. However, the major drawback is that it is not electric or hybrid powered.

PAL-V of Germany offers the Liberty flying car, with the tagline 'the drive to fly', for two people. On land, it has a speed of 160 km/h and a range of 1,315 km/h, transforms into a gyroplane in 5 minutes with a range of up to

500 km/h at 11,000 feet.[23] Standard price is EUR299,000, and further personalization and configuration is provided for a higher fee. To remove challenges such as lack of training, there is a Fly Drive Academy. For corporate businesses or leisure travellers looking for a new skill, if rental were an option, this would be a popular experience and new skill to learn.

Chinese company, XPeng AeroHT, is developing a modular flying car, combining road and air modules that separate and combine. Based in Guangzhou, these vehicles will be integrated in the Greater Bay Area integrated transport system, with the Chinese government aiming to invest US$1.4 billion in the new low-altitude economy.[24] The company has received certification in China in 2023 and in Dubai in 2022, and showcased its flight capabilities at CES 2024 in Las Vegas. Electric powered, the 270° panoramic window is bound to be a hit with travellers for the next generation of air/car travel. Mass production is scheduled for 2025.

Chinese autonomous eVTOL company EHang is also poised to benefit from the low-altitude economy in Guandong province, having already received the relevant certificates for model type and air worthiness. Partnership opportunities are opening up for the next generation of sustainable transport to collaborate with travel businesses.

TIPS

- For innovation to merit funding and investment, it needs to solve the main challenges faced in preserving humanity and the planet for future generations

- If the low-altitude economy is not regulated properly, with excessive numbers of eVTOLs and delivery drones, there could potentially be a consumer backlash.

- Investing in innovative flying cars that are not electric appears counter-intuitive.

- Innovation should be seen through the lens of purpose and not perpetuate inequality.

- For innovation to work, extensive partnerships across academia, private and public sector must be in place.

- Flying cars offer a unique and futuristic experience for travellers looking for a next - generation experience where rental options would be an easier way for access.

Electric vehicles and autonomous vehicles roadmap

On the ground, the future of vehicles is electric. The electric vehicles (EV) boom is in full swing and huge opportunities exist as consumers and businesses make the gradual transition to EVs, spurred on by increased EV models available to buy and incentivized by government policies. Based on announced pledges by manufacturers, the forecast is for 525 million EVs to be on the roads worldwide by 2035, a twelve-fold increase and accounting for 1 in 4 cars.[25] The EV race is led by China far out in front, followed by Europe, with the US trailing behind.

Fransua Vytautas Razvadauskas, Mobility Insights Manager at Euromonitor International, explains that China is leaps and bounds above the rest and it will be difficult to break their dominance as they have a major advantage in the upstream supply process such as metal production, refining, battery production and rare earth metals.

However, the problem with policies is that they can be shelved when there is regime change. Affordability and access to charging infrastructure are key stumbling blocks to uptake of EVs. Consumers experience charging anxiety if the EV infrastructure is not in place to meet their charging needs. This has led to a global slowdown in the transition to EVs as consumer adoption is restricted by obstacles such as availability of charge points and speed of charge. The average ratio of vehicles to charging points is 10:1.[26] The EU, for example, aims for a fast charging point every 60 km on its main highways in the EU TransEuropean Transport Network (TEN-T) in accordance with the Alternative Fuel Infrastructure Regulation (AFIR).

To combat such challenges, for example, the UK government announced in 2024 that it would move quicker and lowered its mandate for 80 per cent of cars and 70 per cent of vans sold in the UK to be zero emissions vehicles (ZEV) by 2030, from 2035.[27] To ensure a just transition that does not exclude lower income families, the UK government is providing financial support, grants and making changes to the law to provide transparent pricing for charge points and contactless payments. In April 2024, the UK charging network included 53,529 workplace charging points compared to only 6,665 domestic devices and 8,354 on-street residential devices, showing that there is a long road ahead.[28] Landlords can benefit from grants to introduce charging infrastructure, and some hotels are already providing charging capacity to guests and staff. Charging EV owners to use charge points creates a new revenue stream which could be replicated for travel businesses.

The competition for leading the net zero emissions race through EVs is intense, led by global brands like Tesla. With car manufacturers' pledges to electrify, up to 58 per cent of cars globally could be electric by 2030.[29] Car manufacturers BYD, SAIC and Geely in China and Tata in India are leading the race.

From fully electric to autonomous electric vehicles will require a step-change in R&D and new technologies like AI and V2X, however, the future is not far off. The forecast value for the autonomous vehicle market is US$2.5 trillion by 2040 for personal and shared vehicles.[30]

However, the Society of Automotive Engineers' (SAE) various levels of autonomous driving need to be moved through first. Mercedes-Benz is the first to receive approval for level 3 conditionally automated driving in the US, in Nevada, as it met the minimal risk requirements. However, its Drive Pilot system means travelling at 40 mph and the driver must be willing to step in and take over if needed.

Tesla's Autopilot and full self-driving capability are a step closer to greater autonomy but still require driver intervention. Some of the self-driving features include smart summon where the car will come and find its driver, with the touch of the app, as well as auto lane change. Tesla's partnership with Baidu in China for 3D mapping may accelerate its entry into the future driverless EV market in China.

AVs are not common, apart from cities like Abu Dhabi (UAE), where a fleet of SAE level 4 TXAI by Bayanat have been in circulation since 2021 on Yas Island and Saadiyat Island. The beauty is that they are electric, free and can be hailed via an app for ease, making them highly inclusive. However, the AV market suffered a major blow due to safety concerns following a Cruise AV that crashed into a pedestrian in San Francisco, leading to the ban of AVs in the city and a suspension of General Motor's Cruise AVs nationwide.

'5G was said to be the enabler of AVs so 6G will provide even better systems with lower latency making it a more attractive option. However, on the other hand, safety is a concern, undermining consumer acceptance and trust whilst regulatory challenges need to be ironed out about responsibility in cases of accident.'

Fransua Vytautas Razvadauskas, Mobility Insights Manager, Euromonitor International

TIPS

- With travel businesses heavily reliant on land transport for transfers, tours and transit, there is a pressing need to electrify fleet globally.

- Travel businesses and destinations should look to expand their charging infrastructure network and solutions, including ensuring contactless as well as cash payments for older generations.

- Opening up charging infrastructure to local communities will boost integration and create potential new revenue streams.

Infrastructure investment and the funding gap challenge

With urbanization continuing into the next century as people continue to move to cities for jobs, education and opportunities, so grows the need for infrastructure to meet their needs. Yet infrastructure is one of the key contributors to carbon emissions especially buildings and construction. Goal 9 of the SDGs relates to industry infrastructure and innovation. To meet this goal, the infrastructure sector is transforming to ensure sustainable and equitable outcomes. Yet this is a major challenge where US$3.3 trillion needs to be spent annually on infrastructure to 2050 to achieve the goals and make countries and cities sustainable, resilient and inclusive.[31]

The so-called infrastructure investment gap is most critical in emerging and developing countries where adaptive capacities are required to deal with the impacts of climate change. However, there is a clear disconnect between national infrastructure development plans and sustainability goals where the latter rarely incorporates the former. With 75 per cent of the world's infrastructure still to be built, there are opportunities as well as challenges to ensure that the future built environment is equally client-resilient, as well as smart, inclusive and climate-positive.[32]

Transport is the largest recipient for investment in the G20, accounting for 42 per cent of the US$1 trillion spending on infrastructure, where the latter accounts for 1 per cent of G20 GDP in 2022.[33] Under transport investment of US$416 billion, the majority of 47 per cent is earmarked for roads, 26 per cent for rail, 10 per cent for public transport, 3 per cent for ZEVs with the remainder accounted for by ports and other types of transport.[34]

The majority of investment comes from government, and more efforts need to be made to scale the required private capital investment to close the infrastructure gap, especially in emerging economies that received only 29 per cent.[35] Tourism, arts and culture receive a mere US$3 billion per annum, amounting to 2 per cent of the US$166 billion spent on social infrastructure.[36] Public private partnerships (PPP) are therefore fundamentally important especially in emerging markets to plug the gap for all.

Infrastructure has been designated as being integral to delivering the following transformative outcomes: resilience, inclusivity, economic development, environmental sustainability, R&D and digital/Infratech.

When it comes to infrastructure for transformation, some countries are streets ahead. Saudi Arabia has even coined the term 'gigaproject' to capture the unprecedented scale of its investments, with 14 giga projects under its Saudi's Vision 2030 plans. The country forecasts that US$7.1 trillion is required to fund turning the vision into reality by 2030, with its Public Investment Fund (PIF) playing a major role.[37] The scale of investment reflects the level of diversification required to transition away from a fossil fuel-driven economy to leisure tourism, sports and culture in addition to its well-established religious tourism. NEOM, one of the flagship gigaprojects, alone is expected to cost US$500 billion, to build one of the world's largest construction projects.[38]

Importance of a sustainable and resilient built environment

Travel businesses and destinations are highly dependent on physical assets such as buildings, and the energy and transport to support them. Construction is the most polluting sector, accounting for 37 per cent of carbon emissions, using carbon intensive materials like cement, steel and aluminium.[39] A whole lifecycle approach is therefore necessary to future-proof construction, architecture and the materials used from the beginning to end of use. Alternative materials that are bio-based are garnering much interest, including wood, timber, bamboo and biomass. Embracing circularity for the built environment is the recommended approach.

Hotels have long been early adopters of Leadership in Energy and Environmental Design (LEED) certified buildings. Airports, museums, convention centres and sports stadia can also be LEED certified as well as cities and communities. The next generation of LEED v5 will encompass decarbonization, quality of life being people-centric as well as environmental and conservation-friendly.

The transformation to net zero is taking shape with the Buildings Breakthrough initiative's goal for sustainable and resilient near net zero buildings to be the new normal by 2030 launched in the UAE at COP28. However, so far the signatories sit at 28 countries led by France and

Morocco, with some notable exceptions such as India, Russia and South Africa, Mexico, Saudi Arabia and the UAE.[40]

As with most certification, there is more than one. An alternative to LEED is BREEAM offering sustainability certification and verification for the built environment and infrastructure on the road to net zero. CitizenM, a trendy lifestyle hotel, works with the BREEAM framework in the US.

The need for sustainable and resilient buildings is global, where regions like Africa will see the largest increase in people by 2050, but 80 per cent of the buildings required have not been built yet.[41] It is equally important to not just focus on net zero but also universal design for the inclusion of all, regardless of ability or age.

TIPS

- With the gap in infrastructure spending and lack of built capacity in emerging economies, there is an opportunity for travel businesses to fund local infrastructure projects that deliver on the SDG and net zero agenda.

- Construction is a major challenge to decarbonization but also opens up opportunities for retrofitting and renovation, breathing new life into the built environment.

- Taking a universal design mindset will ensure inclusivity.

- Certification is the way to go for transparency.

Supersonic renaissance to recreate the golden age of travel, sustainably

The dream of supersonic flight continues to appeal to consumers, investors and innovators, recreated for a net zero world. Concorde, a joint British/French cooperation, remains the iconic supersonic aircraft that travelled at the speed of sound, but beset with high costs and a tragic accident, retired in 2003. With a cruising speed of 1,350 mph/2,160 km/h (mach 2), Concorde travelled at twice the speed of light, cutting journeys in half, with New York to London in under three hours. During its time, Concorde carried over 2.5 million passengers.[42] However, the price was prohibitive.

Companies like Boom in the US aim to recreate the golden age of supersonic travel but in a sustainable way. Yet the sector is nascent with bans on supersonic flights and testing only in the US. Boom is by no means the only contender in the supersonic race with countries like China, Russia and Japan involved and companies like Spike Aerospace.

NASA is making headway with its supersonic passenger aircraft testing with the X-59 experiment in partnership with Lockheed Martin to minimize the sonic boom that would enable flying over land without upsetting communities. The X-59 is due to fly at 1.4 mach (950 mph). Yet success is not guaranteed, with corporate failures like Aerion due to lack of funding.

REAL-WORLD EXAMPLE
Boom aims to redraw the route map with supersonic travel

Supersonic travel continues to capture the imagination of many from business travellers to investors and innovators. Boom is a US private company that is backed by leading tech founders and investors of the online age including Airbnb, Google and Stripe, and groups like Y Combinator, Emerson Collective, Bessemer Venture Partners and Prime Movers Lab.

Its Overture supersonic aircraft has caught the attention of major global airlines, with 130 pre-orders from the likes of American Airlines, United Airlines and Japan Airlines already on the books while it undergoes testing. Its maximum cruise speed is 1.7 mach, at 60,000 feet, with a range of 4,888 miles/7,867 km, with a maximum passenger capacity of 80.[43] The most appealing feature of the Overture, besides speed, is that it is 100 per cent optimized for sustainable aviation fuel. Decarbonization is a fundamental pillar of Boom and its vision for sustainable supersonic air travel, across the entire supply chain from design, manufacturing and flight.

The aim is for net zero carbon by 2025 and net zero GHG emissions by 2040, signing up to the SBTi. It has committed to reduce scope 1 and scope 2 GHG emissions by 42 per cent by 2030 from a 2021 base year, and to measure and reduce its scope 3 emissions.[44] To meet these goals, Boom partners with Watershed for measurement and Frontier for carbon capture for 1,000 years.

SAF is critical to deliver net zero emissions for flight. Working with Sustainable Aviation Buyers Alliance (SABA) enables it to help reduce its own business travel carbon emissions while it scales SAF. Boom is actively engaged in ensuring that aviation can decarbonize and meet its 2050 goals, through scaling up SAF production and it is hopeful that this pathway can be achieved by 2040, if public and private sectors align.

Once supersonic travel returns by 2030, North America will be ever closer to Asia Pacific, Australasia and the Middle East as journey times will dramatically decrease with North America to Asia Pacific times almost halved, opening up new travel corridors. With 600 routes identified, supersonic could reshape where consumers travel in future. The price will be the determining factor for adoption, where airlines will set pricing likely on a par with current business and first-class travel. This will push supersonic travel out of reach of the general public.

For long-haul business travellers, jetting to Paris from the US for the day rather than three days could be beneficial timewise, yet not for the local community. The true sustainability impacts of supersonic travel remain in question. Even with the fate of supersonic flight still unknown, the race for hypersonic flight is already taking place around the world for speeds of over mach 5.

Companies such as Destinus in Europe are testing and developing hypersonic aircraft for 100 passengers, fuelled by hydrogen. It will test the first ever hydrogen-powered hypersonic prototype in 2026 and build a hypersonic airplane by 2030. Due to the sonic boom, such ultra-long-haul hypersonic flights will be restricted to over ocean or desert, so-called bypass routes. The technology is transformative, where hypersonic aircraft are forecast to cover 18,000 km in three hours versus a subsonic aircraft taking 14 hours. The Destinus S aims to revolutionize business travel, with New York to Paris in 1.5 hours, carrying 25 passengers over 10,000 km with deliveries from 2032.[45] Hydrogen will power the turbo jet and ramjet required to achieve hypersonic speeds, where hydrogen will also help cooling which is a challenge for hypersonic as well as supersonic aircraft.

TIPS

- The next generation of supersonic and hypersonic aircraft that are cleaner and faster than the speed of sound will make long-haul destinations ever closer.

- Price will be prohibitive and carbon impacts will be a key determinant of success.

- For the high-end business traveller, celebrities and HNWIs, the future of air travel looks more convenient.

Space travel – unprecedented exploration with challenges to overcome

Space continues to capture the world's imagination, with the opportunities of space travel within closer reach thanks to the efforts of commercial pioneers like Virgin Galactic, Blue Origin and SpaceX.

Virgin Galactic, owned by Richard Branson, has its spaceport in New Mexico. Flights are for 17 passengers, experiencing what is usually the domain of trained astronauts. At a price of US$450,000, the journey into space lasts 90 minutes, at a speed of mach 3.[46] After decades of R&D, Virgin Galactic finally flew its first space tourism flight in 2023 on the VSS Unity, flying three passengers 50 miles/80 km above the earth.

With 12 space flights under its belt with the VSS Unity, Virgin Galactic will launch its new Delta Class aircraft in 2026. The next-generation spacecraft will enable 12 times more frequency of space missions of eight per month.[47]

Branson has competition with other world-famous billionaires and their space companies. The ultimate goal of SpaceX Technologies Corp, owned by Elon Musk, is to take mankind to Mars – one million people with the necessary equipment. Missions that will take people into space include trips in Earth's orbit, the International Space Station (ISS), the Moon and Mars with bookings from late 2024. Missions to Earth's orbit can accommodate two to four passengers, while 12 passengers will travel to the Moon with their own quarters plus a space for socializing.[48] The Dragon spacecraft has capacity for seven passengers, the first privately owned craft to travel to the ISS. In 2024, tests of the Starship craft continued with the aim for the return and reuse of spacecraft and enabling increased frequency of missions. An important feature of SpaceX is that all missions will bring benefits to advancing scientific knowledge.

However, there are major concerns already raised about the environmental impacts of more frequent launches into space and on earth, where space exploration sits outside international climate treaties like the Paris Agreement. Considering that international aviation remains outside of the Paris Agreement, it is not a surprise that space travel is also excluded from climate targets. However, the current status quo needs to be reviewed, considering the carbon footprint of space launch is 100 times more harmful than subsonic air travel due to the release of black carbon with warming effects in the stratosphere.[49]

SpaceX also has plans to transform long-haul travel, with capacity for 100 people on Starship, travelling from one side of the world to the other in less than one hour by entering the Earth's orbit. Greater scrutiny will be needed to regulate Earth–Earth space travel.

Blue Origin owned by Jeff Bezos is involved in space exploration and space tourism with a stronger sustainability approach than its competitors through reusability of engines and rockets. Equally, it aims to use cleaner fuels such as hydrogen with liquid oxygen or liquified natural gas (LNG). It has adopted a circular and sustainable supply chain and operations. The company is equally applying sustainable practices and technologies in space such as carbon neutral, non-toxic and solar-powered. It has made exciting breakthroughs such as Blue Alchemist technology that can transform lunar regolith simulant into solar cells and wires that produce electricity and byproducts such as oxygen.[50] This approach makes the most of the Moon's or Mars's natural resources, without the need to transport materials from Earth, making it more sustainable. This makes life on the Moon and Mars one step more achievable.

Earth observation of carbon emissions and mitigation is called for by leading scientists to meet the Paris Agreement targets, so space has an important role to play in fighting climate change. A balanced view of the cost benefits to ensure sustainability on Earth as well as in space is necessary. The non-binding Artemis Accords provide the framework for the principles of operating in space, promoting peace, transparency, sustainability and interoperability with 42 signatories including the US and noticeable absences like China and Russia. Space debris is a major challenge as well as damage to the ozone layer from spacecraft emissions.

TIPS

- The ventures into space may be out of reach of the general public, but it is important that space travel is accessible for all such as through a lottery system to encourage participation.
- With more space ports popping up, star-gazing experiences and astro-tourism will receive a welcome boost especially where dark skies prosper.

Far-flung destinations like New Zealand, Australia and Japan are investing heavily in their space industries for job creation and to take a slice of the forecast US$1.8 trillion space economy by 2035.[51] Long-haul flights could see an erosion from the emergence of suborbital business travel. As the space economy triples in size between 2023 to 2035, space travel will thrive, but at what cost to the environment and the residents of planet earth?

TIPS

- Regulating space travel and the space economy will create a framework for success taking account of holistic impacts.

- As space travel increases in frequency with more rockets and space debris, a firmer understanding of space exploration companies' carbon footprints will be necessary.

- Replacing long-haul air travel with low Earth suborbital flights that entail a multiple-fold increase in carbon emissions is nonsensical.

- Just because the technology exists, does not mean that it should be used, even if time-saving for the wealthy.

- Alternatively, AR/VR is a cheaper, less carbon-intensive way to experience the Earth's overview.

Next gen maritime innovation and navigation

Air travel whether supersonic, hypersonic, suborbital or interplanetary is not the only mode of transport receiving investment in innovation to meet its net zero ambitions by 2050. Electrification of ships is not taking place at the scale of cars, yet there are moves in that direction using batteries or hydrogen fuel cells for all-electric or hybrid ships.

Norway passed regulations for net zero emission for two of its UNESCO World Heritage fjords to be in place by 2026. To meet this, two of the world's largest hydrogen ferries are being manufactured locally, leading to a new generation of ships, ordered by Torghatten Nord and due in service in 2026.

The maritime sector is pursuing the transitional fuel, LNG, as part of its decarbonization pathway. The different stages involve shifting from LNG to bioLNG which is already being dropped in and eventually e-LNG once available. The sector is looking at ways to meet its regulatory commitments under the EU ETS scheme (2024) and FuelEU Maritime (2025) and net zero emissions by 2050. Scientists dispute the move to LNG is environmentally or financially sound. Instead, the advice is to shift to electric ships for short voyages and hydrogen powered for deep sea voyages.

EXAMPLE
PONANT's next gen vessel

PONANT (France) revealed its next-generation ship, taking a whole lifecycle approach, called SWAP2Zero, signifying sustainable, wind-assisted propulsion and zero emission ready. Due for launch by 2030, SWAP2Zero will usher in a new era of carbon-neutral cruise vessels. With PONANT offering cruise trips all over the world including to fragile regions like Antarctica and the Pacific, its innovative technology will put it ahead of its competitors.

Using a multi-energy approach, the next-generation sailing ship includes hydrogen fuel cell technology, combined with solar and wind energy from its sails and uses direct carbon capture. It is built with slow travel in mind due to its reliance on wind power that will account for half of its energy source. PONANT is aiming to deliver net zero faster than the global target, with 80 per cent reduction by 2040 to meet the 2050 deadline.

With sea levels rising and increased flooding from climate change, a neglected sector of the travel industry, amphibious vehicles, may see revived interest if electrified. The Duck Tour enjoys small success in Gothenburg and Stockholm (Sweden) and select UK cities with Oceanbus. Les Canards de Paris are a fun way to enjoy the city of Paris and the River Seine. However, the sector has witnessed issues with safety from fatal accidents.

Some innovative concepts are emerging providing multi-functional vehicles. BeTriton is an electric bike/amphibious vehicle and camper that can carry and sleep two people. Such innovative niche concepts offer a new way to enjoy sustainable adventure travel.

TIPS

- Some companies will take a slower transition approach to reaching net zero, while others will be bolder and go further, faster.
- Taking a slower and considerate approach will help rediscover the joys of travel.
- Innovative mobility concepts will continue to appeal to younger generations looking for the next new travel experience.

Green breakthrough technologies

Carbon capture and removal

If the world is to succeed in meeting its emissions targets, innovative technologies such as carbon capture, utilization and storage (CCUS) are needed. This is an additional green energy technology that can be further adopted by national governments and industry to meet the net zero pathway. The International Energy Agency reports that only 40 per cent of the carbon required to be captured and removed by 2030 has been achieved to reach the net zero scenario by 2050.[52] Progress is not on track for CCUS, unlike sectors like EVs and solar photovoltaics (PV) that are on track on the clean energy trajectory path to 2030 for 2050.[53] For travel businesses, the links to CCUS are more indirect, where they can opt for energy providers or construction companies that invest in and utilize CCUS. Taking a holistic approach to carbon emissions across scopes 1, 2 and 3 requires taking account of new approaches such as carbon capture.

CCUS differs from carbon offsetting where the former directly removes CO_2 from the atmosphere when fossil fuels are burnt and stores it thousands of feet underground, offering a carbon negative solution. The capture takes place at industrial plants or grids, whereas carbon offsetting involves paying for emissions to be reduced elsewhere by buying carbon credits for offsetting schemes such as reforestation or investing in renewable energy.

Direct air capture (DAC) is expensive and a nascent sector, where giant fans suck CO_2 from the air. One use for the CO_2 captured is to produce cement, synthetic SAF and hydrogen for cleaner transport fuels. The sector received a boost from Blackrock investing US$500 million in Occidental (Oxy), the world's largest DAC company.[54]

TIPS

- One approach to carbon removal is not sufficient.

- An all-encompassing approach is required to use all types including natural capture, direct air capture and other technological techniques.

- Finding like-minded climate-positive partners adopting next-generation energy solutions is key.

REAL-WORLD EXAMPLE
Tomorrow's Air pioneers carbon removal and SAF for the sustainable future of travel

Tomorrow's Air was created by Christina Beckmann within the Adventure Travel Trade Association (ATTA) after an inspiring trip to Antarctica, witnessing the real impacts of climate change. Tomorrow's Air is bringing together passionate travel businesses, travellers and artists united by the desire to help make travel sustainable through emissions reduction and carbon removal with permanent storage. The community engaged with over 77,000 people in 2023, sharing climate-conscious travel tips, inspiring stories and practical information about emerging solutions to help the climate.

Tomorrow's Air offers two ways for individual travellers to engage – by supporting carbon removal with permanent storage and/or sustainable aviation fuel (SAF). In 2023, through partnerships with 17 travel businesses over 23,000 travellers helped order over 300 tonnes of CO_2 removal, double the amount ordered in 2022.

Its carbon removal partners include pure tech and hybrid nature-tech innovations that offer carbon removal with permanent storage: Climeworks, Eion, Octavia Carbon and Pacific Biochar. In 2023 Neste sustainable aviation fuel was also added as an option. Using sustainable aviation fuel (SAF) to replace fossil jet fuel enables a reduction of GHG emissions of up to 80 per cent over the fuel's life cycle compared to using fossil jet fuel.

Among the travel businesses adding Tomorrow's Air as part of their sustainability strategy are kimkim, Natural Habitat Adventures and Geographic Expeditions. In the case of kimkim, a US online travel agency, an additional US$20 per booking is directed to its Climate Fund. Through the Climate Fund, the company invests in local projects and sends a portion of funding to Tomorrow's Air to support climate-conscious travel education, carbon removal with permanent storage and sustainable aviation fuel.

Communication and education are key pillars for Tomorrow's Air to make it easy for travel businesses and their travellers to bridge the intention-action gap. Carbon removal and storage is complex but vital: by 2050, 10 gigatons per year of carbon removal will be necessary. Tomorrow's Air is leading the charge to demystify the science and empower travellers and businesses to support breakthrough innovations.

'I believe travellers can be a powerful force for good in helping spur uptake of solutions that can help us reduce emissions and clean them up.'

Christina Beckmann, Founder, Tomorrow's Air

Summary

Although there are already some signs of what will emerge, the next generation of innovative, sustainable technology remains elusive until 6G arrives. History shows that there will be exponential leaps forward in computational speed, new business models and revenue streams. By 2030, a new era of unprecedented hyperconnectivity and data capture is expected, driving decarbonization, operational efficiencies and personalization that will benefit businesses and consumers.

Against a backdrop of rapid decarbonization, innovative solutions and ideas are required to problem solve and create shared value. Transport is the main culprit for the industry's emissions and is a key target of innovation – some useful, some vanity-driven. Next-generation technologies that embrace decarbonization for a net zero world are garnering more attention from investors and start-ups, while moon-shot projects, such as taking humanity to Mars, also appeal to wealthy investors. Technology should never be used for technology's sake but as an enabler. The primary focus should be on preservation of people, planet and biodiversity where the regenerative power of nature can be deployed to thrive.

However, climate-positive government policies are often fleeting, driven by geopolitical winds. This makes it difficult to deliver on the long-term transformations required for decarbonization as no business operates in silo. A collective system-wide approach is therefore required to deliver on the sustainable development agenda for a fair, inclusive and net zero world. Finding like-minded partners in travel and beyond – mobility, travel tech, technology, energy, construction, food and drinks, public sector, NGOs and academia – will help pave the most resilient pathways forward. However, the choice of different pathways is closing fast as new climate tipping points are breached. Travel businesses should reject short-termism, embrace change and apply a laser-sharp determination to the opportunities and challenges ahead. Agility and adaptability will be key to overcoming obstacles.

As stewards of communities, destinations and biodiversity, travel businesses are in a unique position to lead the necessary transformation for a healthy and prosperous planet, future-proofed for all generations to come.

Endnotes

1   University of Exeter (2023) news.exeter.ac.uk/research/limiting-global-warming-to-1-5c-would-save-billions-from-dangerously-hot-climate/ (archived at https://perma.cc/ARZ6-DH5G)

2   Ray Kurzweil (2007) ntrs.nasa.gov/api/citations/20070038360/
    downloads/20070038360.pdf (archived at https://perma.cc/3QPK-WJ87)

3   Ericsson (2024) www.ericsson.com/en/6g (archived at https://perma.cc/26PW-
    NDL4)

4   Live Science (2024) www.livescience.com/technology/communications/6g-
    speeds-hit-100-gbps-in-new-test-500-times-faster-than-average-5g-
    cellphones#:~:text=6G%20speeds%20hit%20100%20Gbps,average%20
    5G%20cellphones%20%7C%20Live%20Science (archived at https://perma.cc/
    J35W-95VV)

5   IPCC (2023) www.ipcc.ch/report/ar6/wg3/downloads/report/IPCC_AR6_
    WGIII_Chapter10.pdf (archived at https://perma.cc/FPY5-T3AM)

6   Uber (2024) investor.uber.com/news-events/news/press-release-details/2024/
    Uber-Announces-Results-for-Fourth-Quarter-and-Full-Year-2023/default.aspx
    (archived at https://perma.cc/6ZA4-3YH5)

7   National High Speed Rail Corporation Ltd (2024) nhsrcl.in/en/project/
    mahsr-project-operational-planss (archived at https://perma.cc/8URC-PZND)

8   European Energy Agency (2020) www.eea.europa.eu/publications/transport-
    and-environment-report-2020 (archived at https://perma.cc/53T9-4Q7G)

9   Indian Railways (2024) indianrailways.gov.in/NRP%2015th%20DEC.pdf
    (archived at https://perma.cc/FH89-E94B)

10 Ibid

11 Arabian Gulf Business Insights (2023) www.agbi.com/infrastructure/2023/12/
    big-5-construction-gcc-rail-network/ (archived at https://perma.cc/34DV-
    DMQD)

12 Bloomberg (2023) www.bloomberg.com/news/articles/2023-05-12/elon-musk-
    s-hyperloop-vision-gets-a-jolt-of-energy-with-htt-italian-deal (archived at
    https://perma.cc/P8JM-LG3C)

13 PWC (2021) www.strategyand.pwc.com/it/en/assets/pdf/S&-path-towards-a-
    mobility-in-the-third-dimension.pdf (archived at https://perma.cc/F4ZC-8QCA)

14 Nature (2019) www.nature.com/articles/s41467-019-09426-0#Abs1 (archived
    at https://perma.cc/8BYL-BLX7)

15 Volocopter (2019) assets.ctfassets.net/vnrac6vfvrab/73kYdf0o0kR7Y8XqAz9r
    El/40bcf5c38552f6d1fcca71f7fe9736f3/20220607_VoloCity_Specs.pdf
    (archived at https://perma.cc/SAW9-5HAT)

16 Lyte Aviation (2023) lyteaviation.com/la-44/ (archived at https://perma.cc/
    8QPT-E7GN)

17 Eve Air Mobility (2024) ir.eveairmobility.com/sec-filings/annual-reports##
    document-499-0001554855-24-000091-2 (archived at https://perma.cc/
    LCP7-GW6X)

18 Eve Air Mobility (2024) www.eveairmobility.com (archived at https://perma.cc/
    NL62-UL2S)

19 Ibid

20 Eve Air Mobility (2023) www.eveairmobility.com/storage/2023/03/UK_Air_
    Mobility_Consortium_UAM_CONOPS.pdf (archived at https://perma.cc/
    J7S5-PFXL)

21 Mckinsey (2023) www.mckinsey.com/industries/automotive-and-assembly/our-insights/shared-mobility-sustainable-cities-shared-destinies (archived at https://perma.cc/R8F3-PWGB)

22 CNBC (2024) www.cnbc.com/2024/03/04/flying-car-firm-alef-hits-2850-preorders-worth-over-850-million.html#:~:text=BARCELONA%2C%20Spain%20%E2%80%94%20Alef%20Aeronautics%2C,and%20landing%20(eVTOL)%20vehicle (archived at https://perma.cc/9Z6D-DSGV).

23 PAL-V (2024) www.pal-v.com/en/liberty (archived at https://perma.cc/VEF2-LUND)

24 Nikkei Asia (2024) asia.nikkei.com/Business/China-tech/China-s-Guangzhou-to-invest-1.4bn-in-flying-car-infrastructure (archived at https://perma.cc/F2QS-YUTL)

25 International Energy Agency (2024) www.iea.org/reports/global-ev-outlook-2024/outlook-for-electric-mobility (archived at https://perma.cc/5UB3-L774)

26 Euromonitor International from IEA (2024)

27 UK Government (2024) www.gov.uk/government/news/rollout-of-electric-vehicle-chargepoints-to-be-accelerated#:~:text=The%20zero%20emission%20vehicle%20(ZEV,up%20the%20rollout%20of%20chargepoints (archived at https://perma.cc/G43A-RHSQ).

28 UK Department of Transport (2024) www.gov.uk/government/statistics/electric-vehicle-charging-device-grant-scheme-statistics-april-2024/electric-vehicle-charging-device-grant-scheme-statistics-april-2024#headline-figures (archived at https://perma.cc/4FLE-NMA9)

29 International Energy Agency (2024) www.iea.org/reports/global-ev-outlook-2024/outlook-for-electric-mobility (archived at https://perma.cc/26RB-ZEA3)

30 IDTechEx (2021) idtechex.com/en/research-report/autonomous-cars-and-robotaxis-2020-2040-players-technologies-and-market-forecast/701 (archived at https://perma.cc/98WZ-M6BF)

31 Global Infrastructure Facility/World Bank (2022) www.globalinfrafacility.org/sites/gif/files/2022-10/IWG%20Global%20Stocktake%20Report_final_v2.pdf (archived at https://perma.cc/DM79-GGVH)

32 EY (2024) www.ey.com/en_lu/infrastructure/sustainable-infrastructure--a-paradigm-shift-towards-greener-inv#:~:text=With%20approximately%2075%25%20of%20the,EUR%2010%2C000%20billion%20by%202030 (archived at https://perma.cc/GL89-5GPZ).

33 Global Infrastructure Hub (2024) infratracker.gihub.org/ (archived at https://perma.cc/K9T9-SN9T)

34 Ibid

35 Global Infrastructure Hub (2024) www.gihub.org/infratracker-insights/how-much-do-g20-governments-budget-for-investment-in-infrastructure-annually/ (archived at https://perma.cc/Z6H7-AXJH)

36 Ibid

37 Kingdom of Saudi Arabia (2024) www.vision2030.gov.sa/en/explore-more/national-investment-strategy/ (archived at https://perma.cc/G65F-S7TT)

38 NEOM (2019) www.neom.com/en-us/about (archived at https://perma.cc/85RX-44FD)

39 UN Environment Programme (2023) www.unep.org/resources/report/building-materials-and-climate-constructing-new-future#:~:text=The%20buildings%20and%20construction%20sector,staggering%2037%25%20of%20global%20emissions (archived at https://perma.cc/8QCD-ZTJH).

40 Green Building Advisor (2023) www.greenbuildingadvisor.com/article/28-countries-sign-buildings-breakthrough-agreement-at-cop28 (archived at https://perma.cc/4UQ3-CN8T)

41 World Green Building Council (2024) worldgbc.org/worldgbc-africa-manifesto/ (archived at https://perma.cc/42V3-X2CL)

42 Ibid

43 Boom (2024) boomsupersonic.com/overture (archived at https://perma.cc/NEJ9-TMW6)

44 SBTi (2024) sciencebasedtargets.org/ (archived at https://perma.cc/266E-UHYK)

45 Destinus (2024) www.destinus.com/page/destinus-s (archived at https://perma.cc/69UT-NQQ6)

46 Virgin Galactic (2022) www.virgingalactic.com/news/virgin-galactic-launches-spaceflight-reservations-and-new-consumer-brand (archived at https://perma.cc/Z5DL-8N4G)

47 Ibid

48 Ibid

49 Christopher M Maloney, Robert W Portmann, Martin N Ross, Karen H Rosenlof, (2022) The climate and ozone impacts of black carbon emissions from global rocket launches, *Journal of Geophysical Research (Atmospheres)*, 127 (12).

50 Blue Origin (2024) www.blueorigin.com/news/blue-alchemist-powers-our-lunar-future (archived at https://perma.cc/RHY8-DVN2)

51 World Economic Forum/Mckinsey (2024) www3.weforum.org/docs/WEF_Space_2024.pdf (archived at https://perma.cc/2J4F-XL46)

52 International Energy Agency (2024) www.iea.org/energy-system/carbon-capture-utilisation-and-storage (archived at https://perma.cc/SD7B-JCHE)

53 IEA (2023) www.iea.org/reports/tracking-clean-energy-progress-2023 (archived at https://perma.cc/2NBA-KZT9)

54 Reuters (2023) www.reuters.com/sustainability/climate-energy/blackrock-invest-550-mln-occidentals-carbon-capture-project-2023-11-07/ (archived at https://perma.cc/6HUK-W2KT)

06

Pathways to resilience and purpose

Introduction

As stewards of the environment, biodiversity, culture and communities, travel businesses and destinations are fundamental to creating the resilience needed to pave the way for truly sustainable transformation. Resilience is the ability to deal with change or a crisis, adapting and finding ways to survive and ultimately thrive. Without strong foundations, it is impossible to move forward with transformation in any form, as the mindset and strategies to cope with change and shocks are not in place. Resilience and sustainability go hand in hand, perpetuating a positive feedback loop, required to achieve the Sustainable Development Goals (SDGs). Without resilience there is no sustainable development. There is no resilience without adaptation and mitigation to climate change and reversing biodiversity loss.

Travel businesses of all sizes and destinations know crisis first-hand, facing their worst fear, when all non-essential travel was shut down at the height of the pandemic. Travel businesses, destinations and communities showed true grit in the face of the worst global crisis since World War II. Empowered by the collective will to overcome the challenges, stakeholders came together to pave the way forward, promising no more business as usual.

Post-Covid, new records for peak visitor numbers have been broken for global travel and tourism demand. Normal service has resumed as the industry bounces back to its original form as 'an economic powerhouse' and 'job creation juggernaut'.[1] Enormous resilience was demonstrated in the face of extreme challenges. Yet the bounce back jars with the promises made of transforming to a value-driven tourism model, guided by purpose not just growth.

Going forward, the travel industry faces its greatest existential threat – the triple planetary crisis of climate change, pollution and biodiversity loss. No one company, country or industry can tackle the multiple challenges alone. A system-wide, transformative approach is essential to tackle the threats, involving multiple stakeholders, interdisciplinary approaches and innovative solutions. Adopting circularity in terms of reduce, reuse and recycle principles are fundamental building blocks for success. The climate and biodiversity crises are interlinked and need tackling together.

As Professor of Disruption, Ian Yeoman, of the Hotel Management School Leeuwarden at NHL Stenden University of Applied Sciences says, there are multiple pathways and it is up to us to decide which pathways to choose for shared success.

Travel businesses and governments already know the best pathway, as laid out by the Paris Agreement, is making sure that global warming is no more than 1.5°C of pre-industrial levels. This is the preferred shared socio-economic pathway for sustainable development, as recommended by the InterGovernmental Panel on Climate Change (IPCC), one of their five future scenarios. However, governments' nationally determined commitments (NDCs) and businesses' climate commitments do not go far or fast enough.

Many travel businesses continue to opt for low-hanging fruit, rather than taking a long-term view to wholesale transformation. This short-termism further amplifies future threats and risks. Without changing the perspective to think first and foremost about what local destinations and communities need to thrive, all future action will be in vain.

Building resilience

Frameworks in place to prepare for threats

There is much written in academia and business management about tourism resilience frameworks to prepare for external shocks, threats and uncertainties and how to act accordingly. The complexity for the travel industry is that it is dependent on both the public and private sectors, at the mercy of their respective resilience capabilities. The overarching pillars for future resilience as defined by the World Economic Forum are foresight and preparation, execution and adaptation capabilities.[2] The interconnections across the supply chain complicate the required responses due to the different needs of private/public and the innate fragility in the chain. At the time of disaster, these weaknesses are often exposed, leaving local communities facing the worst impacts of a crisis, often alone in the first few days.

The range of threats that travel businesses and destinations face are varied in terms of type and scale, from natural disasters, terrorism, climatic events, cyber-attacks and war. Future threats, such as the long-term impact of climate change, are often overshadowed by short-term challenges such as geopolitical tensions like East versus West, culture wars or the cost of living crisis.

The 17 SDGs with their 169 targets are the most universal framework that governments and the travel industry are aiming to meet for a sustainable and resilient future. The circular economy is baked into sustainable development so that resources are no longer exploited, they are regenerated. Taking a proactive approach to being climate-ready can be accelerated through a combined approach of the latest technology and nature-positive thinking inspired by indigenous knowledge. With leaps forward in big data, analytics and AI, technology is a powerful evolving tool for destination resilience.

In future, wars and conflict will increasingly be caused by climate-related impacts and pressures, as extreme heat pushes communities out of their homes and land. To minimize climate impacts is a means of safeguarding peace and stability.

RESILIENCE KEY FOCUS AREAS FOR TRAVEL

- Environment and regeneration
- Infrastructure and transport
- Energy
- Water and waste
- Biodiversity and conservation on land and water
- Food security
- Communities and capacity building
- Education and workforce

TIPS

- Risk assessment: frequent reviews of short- and long-term risks will help to be more agile when shocks happen.
- Conducting double materiality assessments is becoming mandatory in regions like Europe which could set a precedent worldwide for ESG reporting.
- Scenario planning is fundamental to forecasting demand and supply, to form adaptive strategies, and put capacity-building measures in place sooner rather than later.
- Data should be central to business decision-making.

Environmental resilience

Some of the most vulnerable countries to climate change and extinction from climate impacts are among the poorest and least developed in the world in regions like sub-Saharan Africa, Latin America and Pacific Islands. Large populous nations such as India, Nigeria, Pakistan, Indonesia and the Philippines are among the most vulnerable to extreme heat by 2100 if global climate targets are not met. Much work is being done on understanding the interlinkages between climate impacts when tipping points are breached and change becomes irreversible.

The IPCC climate change scenarios by range of global warming outline how the higher the temperature increase, the more precipitation there will be for some parts of the world such as Africa, yet greater heat in the poles. Ocean and coastal acidification is not just a threat to ecosystems but to the entire food chain. Every destination whether advanced, emerging, mountain, beach, forest, savannah, urban or otherwise will face its own unique climate-related impacts on its people, biodiversity and environment. Climate change is multifaceted and cost-heavy but, if not invested in, will cost even more to deal with.

REAL-WORLD EXAMPLE

Maldives adapts – using digital innovation for survival with the creation of smart reefs

Many headlines have talked about the potential demise of the Maldives from sea waters rising due to global warming. Despite progress more action is required to adapt and develop resilience, where the country is the 31st most vulnerable country but the 81st ready country.[3] The Maldives is the flattest country in the world, where 80 per cent of its islands are less than 1 metre above mean sea level at risk of floods and storms along with coastal erosion.[4] Its dependency on tourism makes it vulnerable to the combined effects of climate change and the net zero transition, especially if flight prices increase to compensate for emissions. It is at real risk of extinction.

Besides diversifying the tourism offer away from luxury island resorts through encouraging community tourism projects to flourish, technology is critical in data capture to protect not just biodiversity and the environment but the country's future. 'Smart reefs' have been created such as using drones for remote observation to monitor the health of the reef and its biodiversity along with monitoring boat traffic and diving impacts.

The Digital Maldives project is a cross-sector approach to data sharing, and capacity building includes indigenous communities, tourism and fishing businesses supported by the World Bank and Government of the Maldives. Such projects require investment, where US$10 million has been provided by the World Bank for leveraging data and digital technology for climate adaptation and resilience as part of a broader initiative to drive digital connectivity in the country.[5] Building an innovative data platform focused on emissions tracking and climate resilience is a key weapon in protecting the Maldives, its people and prosperity.

Accelerating digitalization equally has helped the activities of conservation NGOs such as Maldives Whaleshark Research Programme that launched a citizen science app for monitoring wildlife through its BigFishNetwork. Hotels like Centara engage their guests in coral planting activities.

Around the world there are coastal areas of countries and cities from Thailand and Pacific Islands to Miami that are at risk of sea levels rising and storm flooding. The level of severity of climate change is dictated by the level of carbon emissions leading to different responses for each global warming pathway based on mean temperature rises compared to pre-industrial levels. With targets aimed at 1.5°C, anything over that, especially 4°C, will lead to even greater flooding and hundreds of millions at risk in coastal areas, among other dire consequences. If, however, targets are met, exposure will be halved and cities will not need to be as shored up or even abandoned.[6]

The list of cities at risk of being underwater include popular destinations such as Venice, Amsterdam, Lisbon, London, Dubai, Bangkok, Ho Chi Min City, Sydney, Tokyo, Mumbai, Istanbul and New York. It is not just people's homes but also important landmarks and transport hubs such as airports that will be submerged. This could mean that iconic sites such as the Statue of Liberty or the Sydney Opera House will disappear under the water. Some cities like Jakarta are even upping sticks and building a new capital city, Nusantara, in Indonesian Borneo or Cairo's new administrative capital in Egypt.

Florida in the US is one of the most at-risk states from flooding and extreme weather events. The state is working to adapt and mitigate the effects of rising sea levels, dependent on increased global temperatures due to carbon emissions and ice sheet melting in places like Greenland and the Antarctic. Focusing on innovative nature-based experiences, leveraging AI for smart marina management and boat traffic navigation are some of the solutions being proposed.

Nature-based solutions to protect fragile destinations

Protecting natural ecosystems, biodiversity and natural defences are integral to resilience. The natural world is facing catastrophe unless there is a reversal in nature losses to nature positive by 2030 and full recovery by 2050. Regions like Latin America, the Caribbean and Africa are witnessing the most significant declines.

Blue carbon ecosystems such as mangroves, sea grass and tidal marshes are highly effective for carbon capture and resilience. Mangroves are blue carbon sinks, better at capturing carbon than even rain forests, at four times more efficient. The Nature Conservancy calls them one of the most important ecosystems, not just protecting against coastal erosion but providing food and income for communities. Mangrove Alliance has developed a significant platform to help with mangroves protection and restoration using geospatial data and remote sensors for monitoring through the Global Mangrove Watch.

The World Wildlife Fund (WWF) is working with four countries around the world to build resilience for communities and their environment through mangroves restoration: Colombia, Mexico, Madagascar and Fiji. The WWF's work takes a holistic approach to environmental and social resilience, increasing carbon capture while opening up new revenue opportunities for communities and MSMEs such as sustainable tourism and beekeeping.

With travellers always on the search for unique and authentic experiences, the prioritization of mangroves opens up new ways for visitors to help drive positive impacts. Funding mechanisms such as incentives help to ensure the long-term success of such projects.

Global hotel chain Iberostar uses nature-based solutions to help with its climate targets and boost destination resilience. It introduced its mangroves programme in 2021, planting over 16,100 mangroves in the Dominican Republic since then. Through its Wave of Change platform and its Coastal Health Roadmap, Iberostar is leading the charge when it comes to nature-based solutions, circularity and decarbonization by 2030. Part of the ocean and coastal actions are dune preservation in Mexico and restoration of coral reefs in the Caribbean. The company also promotes responsible 'blue food' in line with the Blue Food Assessment that includes aquatic animals, algae and plants for responsible sourcing, where 83 per cent of its seafood was responsible in 2023.[7]

With the potential of sea kelp and seaweed for carbon sequestration only beginning to be understood, there is still much to explore in the blue carbon economy. Countries such as Norway are already testing seaweed farming. However, it is never an 'either, or' approach when it comes to decarbonization of areas such as energy, transport and societies in general. Marine conservation is one area that is popular for voluntourism (volunteering) as offered by the likes of travel brands such as Global Nomadic.

TIPS

- Investing in nature-based solutions and the circular economy are integral to regenerative tourism to allow ecosystems to thrive.

- Mangroves are the backdrop to visitor activities from boat tours to wildlife watching so there is an opportunity to promote their importance in the fight against climate change.

- Adventure travel companies are primed to incorporate more conservation-related itineraries.

- There is huge opportunity to transition away from fly and flop holidays to trips that offer giving back experiences, such as beach clean-ups or conservation, especially for families.

- For travel businesses seeking a global standard, the International Union for Conservation of Nature's (IUCN) Nature Based Solutions self-assessment tool is worth considering.

Nature-positive travel

In terms of sustainability, biodiversity often plays second fiddle to the more headline-grabbing climate change. Yet the climate emergency and biodiversity are inextricably linked. Kunming-Montreal Global Biodiversity Framework (GBF) provides the parameters for what success should look like by 2030 and 2050 for living in harmony with nature, as agreed at COP15. By 2030, governments have promised to stop deforestation, stop and reverse nature loss and protect 30 per cent of land and sea marking a major step forward in biodiversity protection. The aim is to bend the curve so that nature can be restored and regenerated so that it can flourish for the benefit of all (see Figure 6.1).

The long-term vision to 2050 has four pillars: protect and restore, prosper with nature, share benefits equally and invest and collaborate. The latter will be challenging as there is a US$700 billion funding gap per year.[8] Central tenets to the Biodiversity Plan for life on earth are the framework for measuring success, transparency and accountability of national plans and the importance of education to ensure sustainable lifestyles compatible with nature.

FIGURE 6.1 Nature positive by 2030

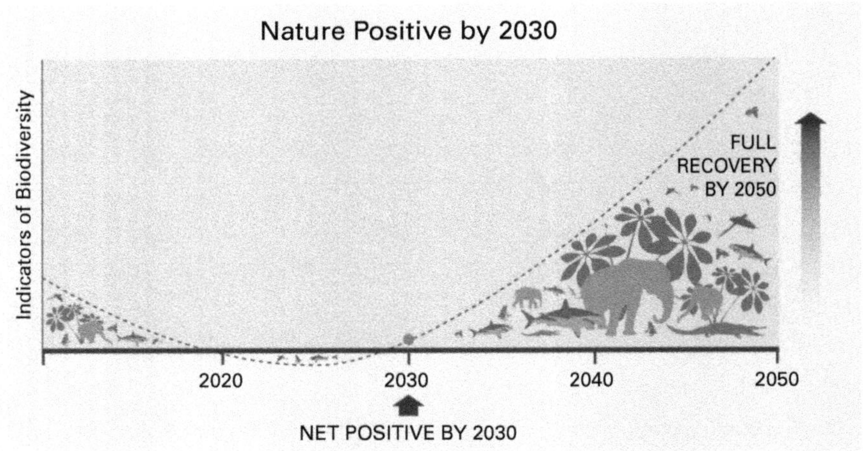

SOURCE Nature Positive Initiative

The sense of urgency to protect biodiversity is acute with scientists saying that we have entered the Sixth Mass Extinction caused by humanity. The WWF's Global Living Planet Index reveals the extent of the damage where there was a decline of 69 per cent over 1970–2018 in monitored populations representing over 5,230 species.[9] Latin America is the worst affected for loss in population abundance (–94 per cent) and freshwater species (–84 per cent).[10] These figures are a resounding clarion call for action.

Living coral reefs have halved in the past 150 years; even with 1.5°C global warming, 70 per cent to 90 per cent of coral reefs will die, and 99 per cent at even 2°C.[11] At 3°C, 41 per cent of mammals will lose their natural habitats, compared to only 4 per cent if the world achieves 1.5°C global warming.[12] The alarming loss of biodiversity has led to a sharper focus on elevating the cause of nature-positive tourism. Interactions with nature can be life-changing, bringing a new sense of value and purpose. As renowned conservationist Dr Jane Goodall said, 'every individual matters, every individual has a role to play, every individual makes a difference'.[13] The interest from travellers is strong, and many are willing to pay more to enjoy nature-based experiences and outdoor activities (see Figure 6.2). Considering the extremity of the crisis in the natural world, there is an opportunity to accelerate more nature-positive experiences especially involving kids.

FIGURE 6.2 Traveller segments interest in paying over 10% for nature and outdoors travel features 2024

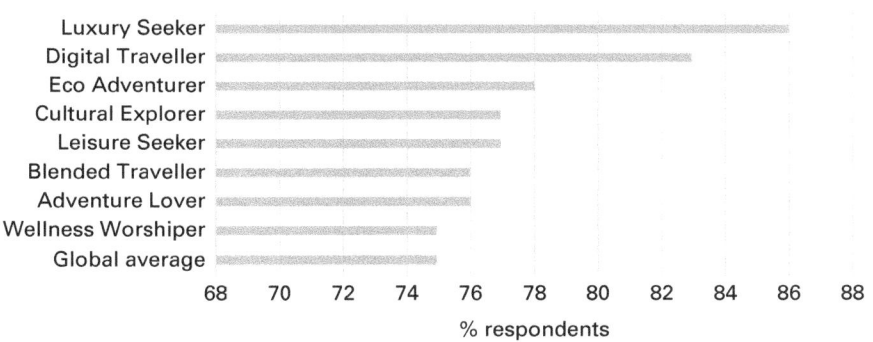

SOURCE Euromonitor International – Voice of the Consumer: Travel Survey
NOTE Consumer respondents n=40,236, fielded in January–February 2024

REAL-WORLD EXAMPLE
Animondial

Animondial is a consultancy specializing and working with travel and tourism businesses to protect nature, stop biodiversity loss and work to transform to a nature-positive industry. In collaboration with WTTC, UN Tourism and Sustainable Hotel Alliance, Animondial and partners launched the Nature-Positive Tourism Partnership at COP15, the UN Convention on Biological Diversity (CBD).

The partnership aims to put travel businesses at the heart of creating nature-positive experiences since the industry is so heavily reliant on the natural world. Travel is only one of six industry sectors globally where 80 per cent of its goods and services are reliant on nature.[14] The mid-term goal is to make travel nature-positive by 2030 as a 'Guardian of Nature'. The partnership follows the Kunming-Montreal Global Biodiversity Framework including protecting and restoring nature along with monitoring impacts on biodiversity and the environment.

Animondial has created Natour, an innovative tool to establish impacts on nature and biodiversity for assessment and measurement, which is aligned with global standards for ESG reporting in line with the SDGs and GSTC. Its Animal Protection Network is another useful resource to find like-minded partners working with organizations such as Ape Africa and Barbados Environmental Conservation Trust. Visitors can get involved with the protection of sea turtles in Costa Rica, for example.

The organization is driven by science. Its central pillars focus on enabling visitors to reconnect with nature, safeguard welfare for nature's right to a quality life, protect endangered species especially from trafficking, invest in nature and stop the sale of degrading experiences with animals. The organization partners with ecollective on carbon impact and the ATTA for delivering nature-positive trainings to adventure travel businesses.

Central pillars for Animondial are identifying negative impacts on nature and tackling challenges such as pollution, land use, climate change and invasive species through nature-based solutions.

The Intergovernmental Science-Policy Platform on Biodiversity and Ecosystem Services (IPEBS) and Intergovernmental Panel on Climate Change (IPCC) presented how biodiversity and climate change are interlinked and should be considered together as both are impacted by human activities. This is due to the complexity of linkages and interconnections, so that work should not be done in isolation in case of unintended consequences. By 2050 climate change will not just be a contributing cause of diversity loss but will be the leading cause, overtaking changes in land-use such as cutting down trees to make way for agriculture.[15]

Reversing climate change and protecting marshlands and mangroves, and planting trees creates a carbon sink to absorb CO_2, but also an ecosystem for wildlife and nature, while preserving coral reefs protects the land and people from storms and coastal erosion and provides food and income from tourism, supporting 1 billion people.[16] With more and more species appearing on the IUCN Red List, 44,016 reported in 2023 and almost 4,000 endangered, there is no time for complacency.[17]

Costa Rica is a world-leading pioneer in conservation and reversing deforestation through its Payments for Environmental Services programme (PES), doubling its rainforest cover in two decades after years of degradation and logging. PES provides financial incentives to communities to reforest and protect biodiversity where tourism has played a major role, with visitors enjoying the abundant rainforest, mangroves and national parks.

TIPS

- Biodiversity is a foundational element of travel from orangutans in Borneo to gorillas in Rwanda where travel businesses and destinations have a profound duty of care to protect the living world.

- It is essential to avoid commoditizing nature-based tourism to ensure that every trip contains educational and interactive elements to inspire biodiversity ambassadors.

- Organizations like the EU are beginning to incorporate nature into their macro-economic modelling through indicators such as the Global Ecosystem Product, going beyond GDP.

- Nature is making its way to being represented at company level – nature on the board.

- In future, voluntary 'climate service' to leverage a pool of citizens may become de rigueur to ensure destinations can proactively deal with climate-related challenges.

Rewilding for regeneration and thriving destinations

In regions like Europe, innovative rewilding is being used to restore natural habitats for lost biodiversity. Innovative projects include the WeWilder Campus that has introduced bison to Romania's Tarcu

Mountains, creating an eco-tourism, workshops and retreat space for visitors and locals. The business supports 20 local families. It is part of Rewilding Europe which has 10 projects from the Danube Delta to Swedish Lapland. Such schemes are helping to reverse wildlife loss, helping nature to make a comeback. An important pillar for rewilding is that nature is in control, allowing natural systems to thrive and build resilience. The priority is that nature and people co-exist in harmony. In Glen Affric (Highlands of Scotland), the Dundreggan Rewilding Centre is the first of its kind, educating visitors and creating a nature-based business for the local community with 15 jobs. It offers immersive nature-based experiences including wellness and equally promotes the Gaelic culture and language (see Figure 6.3).

Fundación Chile Rewilding was inspired by Tomkins Conservation and has created seven national parks, protects 11 million hectares of land and eight endangered species like the puma in Chilean Patagonia.[18] Fundación Rewilding Argentina works in Patagonia and has a notable project in Iberá Park, Corrientes, to protect wetlands in the northeast of the country. Protected animals include jaguar, a major pull for wildlife lovers. It's the country's main nature destination and part of the transnational Litoral Ecotourism corridor.

FIGURE 6.3 Trees for Life – Dundreggan Rewilding Centre

SOURCE Paul Campbell Photography

Conservation tourism – viable alternative to biodiversity loss

Despite the devastating collapse in revenues during the pandemic, tourism continues to be a vehicle for conservation, vital in emerging countries around the world. Wildlife and conservation tourism continues to appeal to adventure travellers, families and those seeking to go off the beaten track. There are multiple luxury travel and specialist players operating in the nature/wildlife space including safari, whale-watching, tiger or polar bear watching. Conservation, restoration and regeneration have become inseparable from community resilience, biodiversity and environmental protection.

Over the past decade, more research has been done on the value of natural capital. The lifetime value of an elephant dead or alive is reported as US$40,000 versus US$1.75 million respectively in terms of its climate-positive activities and carbon capture potential.[19] The value of an individual whale is even greater.

EXAMPLE

The Nature Conservancy champions nature solutions

Global organizations like The Nature Conservancy (TNC) lead the way with their aim to accelerate conservation around the world through natural climate solutions. From the US, Africa and Asia Pacific to China, TNC is championing decarbonization and empowering local and indigenous communities, with creative, innovative and scientific approaches. Volunteers are critical to achieve their 2030 targets, with opportunities to contribute to scientific research, restoration and urban conservation through its Nature Allies programme. The targets include conservation of 10 billion acres of ocean, 1.6 billion acres of land, 620,000 miles of rivers, and reduction or restore 3 gigatons CO_2 per year.[20] Additional goals include helping 100 million people that are exposed to climate risks through restoring ecosystems and empowering local communities and Indigenous People by creating 46 million local stewards.

In some parts of the world like Africa and Asia Pacific, conservation tourism is the lifeblood of communities and biodiversity and stops communities turning to poaching as the only other alternative revenue stream. Australia is also looking to promote safari tourism.

The population of Africa is set to double to 2.4 billion by 2050, and quadruple to 4 billion by 2100 entailing the need for investment in infrastructure, education and healthcare.[21] Travel and tourism can help to drive sustainable and inclusive growth. For visitors, Africa is a dream continent for wildlife from birdwatching to tracking the Big Five, attracting millions of visitors to explore and marvel at wildlife up close. There is a network of transboundary national parks including Peace Parks, private game reserves, safari operators, lodges and the communities themselves. Nevertheless, co-existence with wildlife is challenging and often there is conflict.

Kavango Zambezi (KAZA) trans-frontier conversation area spans five countries – Angola, Botswana, Namibia, Zambia and Zimbabwe – and launched a new sustainable tourism brand, Rivers of Life, in 2024. As a Peace Park, KAZA is as impressive in its size as in its biodiversity including 19 national parks and home to half of the African savannah elephants. It has benefited from a tourist visa scheme for many years.

There is, however, much more to explore in Africa than just safari from cultural, music, sports and adventure to religious tourism, and that is just scratching the surface.

TIPS

- The rapid urbanization in Africa's cities opens up new opportunities for domestic, intra-regional and international tourism that is urban, focused on cultural immersion.

- Promoting travel businesses and visitor economies in Africa's up and coming city destinations should be viewed as a priority, not an afterthought.

REAL-WORLD EXAMPLE

Wilderness Travel – putting positive impact and conservation ahead of profit by closing the loop

Safari with its roots in colonialism and trophy hunting is transforming into a conservation-driven business. Tour operators like Wilderness Travel put positive impact above profit, where their number one goal is the protection, conservation and regeneration of wilderness where biodiversity and communities can thrive. The company owns 2.4 million hectares of private land and aims to double this as its primary business goal is to protect conservation. Around 90 per cent of its employees are from communities local to the parks.

Its impact strategy starts with its camps and concessions, promoting biodiversity health and expanding concessions. This extends to promoting co-existence between communities and wildlife and education, influencing policy and planning at a country level to help secure investment. Finally, there are benefits on a global level by protecting carbon sinks.

In Botswana, Wilderness aimed to solve the problem where there was conflict between elephants and farmers by understanding the movement of the animals, and helping move farmers off animal corridors. This resulted in improved yields and excess millet is used to make local Okavango craft beer. The company buys the local beer to sell in the camps to guests. Essentially the loop has been closed: there is harmony with wildlife and greater food security thanks to market mechanisms.

Fundamental pillars of the company are empower, educate and protect. Eco-camps for kids are offered, but often it's more about providing the resources that they need such as food and school infrastructure. Every year, for five days, the camp is given over to local children to enjoy what visitors experience and change their perspective on wildlife.

In Rwanda, the company promotes reforestation and rewilding, where local communities earn revenue from growing tree nurseries, creating opportunities for SMEs.

'For large, interconnected landscapes to work for wildlife, you need to put all of your efforts into support and management of people. Local community members need to be incentivized through jobs, the purchasing of local skills and products and rewarding people for keeping wildlife alive. This all requires a long-term view and approach if you want to see genuine conservation impact.'

Vincent Shanks, Impact Manager, Wilderness

Role of renewable energy in fostering resilience

Saving energy is a critical concern especially of hotels, where Greenview reported 99 per cent of hotels implemented energy savings over 2020–2022 to help with decarbonization.[22] The transition to renewables is taking place, especially in countries that benefit from natural resources such as solar energy. From Hawaii to Sri Lanka, hotels are investing in photovoltaic (PV) panels. JetWing (Sri Lanka) invested US$1 million in solar power, with additional supply provided to the local grid. In the Maldives, there has been a programme with the World Bank to roll out solar power, where there are even floating solar panels manufactured by Swimsol for resorts. The shift to renewable energy is a key lever to reducing imports of fossil fuels and building stronger, more resilient communities. The Maldives is part of the broader SIDS Lighthouse Initiative, with the approach rolled out to other countries like Seychelles and Mauritius.

Off-grid properties have been a common feature of destinations such as Patagonia, Latin America and Asia Pacific. Being grid-tied is one step better than being self-sufficient, giving back and ensuring a secure energy supply in case of emergency power cuts and blackouts.

In Africa, safari eco lodges often are self-powered by renewables. Some examples are members of The Long Run including Sirikoi (Kenya) and Chumbi Island (Tanzania), while Groobos Private Nature Reserve (South Africa) is a carbon sink and has enjoyed being carbon negative since 2018.

Global Himalayan Expedition (GHE) is an award-winning positive impact travel business that takes visitors into villages without electricity in the Himalayas to install the necessary equipment. This has helped to build 205 resilient communities, positively impacting the lives of 130,000 people while offsetting 120,000 tons of carbon and empowering not just the residents but 1,300 adventure travellers.[23] These expeditions require a joined-up approach from energy companies to NGOs, with remarkable results. Such direct, net positive action by GHE is a real game-changer and a blueprint for travel businesses to emulate in other regions of the world.

Even travel hubs like airports are embracing renewables. Pioneers such as Schiphol Airport (Netherlands) aims to be fully circular and energy positive by 2050, giving back energy to consumers and the grid. Electricity will power vehicles and aircraft. Renewable energy will come from wind, solar and thermal. Every year, the airport makes energy savings of 4 per cent and has been awarded the 4+ accreditation by Airports Council International's Airport Carbon Accreditation scheme.[24]

The leading countries for renewable energy capacity are led by China, the US and Brazil. Surprisingly for hotels, creating or buying in renewable energy is still an uncommon practice, with only less than a quarter using renewables while the majority are focused on energy cost efficiencies.[25]

FIGURE 6.4 Top 10 countries for renewable energy and electricity capacity 2023

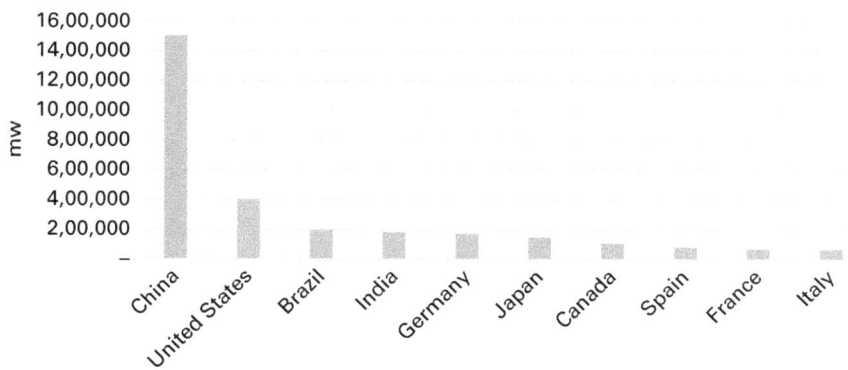

SOURCE International Renewable Energy Agency (IRENA©)[26]
NOTE Includes bioenergy, geothermal, hydropower, marine energy, solar energy, wind energy, pumped storage

---

TIPS

- Travel businesses that provide energy to the grid – net positive style – will be industry leaders.

- Electrification in regions such as Africa is challenging and needs to be accelerated, existing players and future investors should look to convert to net positive renewable energy.

- Energy positive is the gold standard for all travel businesses.

- Travel businesses need to be engaged with the public sector to shape and push for a swifter energy transformation.

Zero plastic for zero waste

Plastic waste and pollution are the scourge of the 21st century, with the Global Commitment launched by the UN Environment Programme (UNEP) and the Ellen MacArthur Foundation in 2018. Progress is not happening fast enough to meet the 2025 targets and there is still a long way to go in achieving zero plastic despite high awareness.

Plastic production is highly carbon intensive using fossil fuels, lethal for biodiversity in the oceans, and for humanity where microplastics have entered the food chain. It pollutes and destroys marine ecosystems. For endangered species like turtles, 22 per cent will die if they eat plastic and 52 per cent of the creatures were found to have ingested plastic according to the WWF.[27]

The Green Lodging Trends Report 2022 reveals that 66 per cent of hotels have removed plastic straws, however, only a mere 7.5 per cent have fully eliminated single use plastics, with a long way to go to eradicate the problem.[28] The majority of hotels are clinging to traditional practices such as the use of plastic water bottles and mini toiletries (see Figure 6.5).

UN Tourism and UNEP launched the Global Tourism Plastics Initiative, with over 200 signatories such as Accor, Booking Holdings and countries like Costa Rica with the aim to reduce, reuse and recycle, particularly to eliminate single use plastic (SUP). The EU is also taking steps to ban SUP through the EU Directive to protect beaches and coastal areas.

The Galapagos Islands (Ecuador) in the Pacific set up the Galapagos Conservation Trust to use innovative solutions to protect its precious and unique biodiversity. The Marine Reserve is a UNESCO-protected site and one of the world's richest natural gems. The programme to return the islands to plastic-free embraces different scientific disciplines and works hand in hand with local communities to design solutions. The aim is to replicate successful approaches in other plastic-ridden areas globally such as Easter Island.

FIGURE 6.5 Hotels' sustainability practices

% respondents

Common	Have at least one initiative to reduce inequalities	97.9
	Implement a linen reuse programme	95.5
	Plan and implement initiatives to reduce energy use	82.4
	Offer guests opportunities to support/participate in sustainability initiatives	78.7
Established	Eliminated plastic straws	66.0
	Plan and implement waste reduction initiatives	62.6
	Track energy and water consumption	60.0
Emerging	Plan and implement water-savings initiatives	49.9
	Replaced mini plastic toiletries	46.6
	Measure carbon emissions	43.1
	Provide vegetarian options	34.5
	Conducted a waste audit in past 3 years	27.3

SOURCE Green Lodging Trends Report 2022, https://greenview.sg/services/green-lodging-trends-report/

NOTE Refers to all hotels surveyed, responses from about 27,000 hotels

TIPS

- Travel businesses should move ahead of compliance when it comes to plastics.
- In the same way that no-fly is increasingly a lifestyle choice, so is no plastic – travel businesses need to be aligned with consumers' values or lose out to circular travel businesses.
- Zero plastic is not new, yet there is a long road ahead requiring more vigilance where it is one of the easier actions to take.
- Bio alternatives to plastics, for example bamboo, beeswax or hemp, are a much more appealing alternative, delivering a more authentic experience while helping the planet.

Reversing deforestation

Planting trees when visiting a destination has long been a popular activity for visitors and dignitaries alike. A more holistic approach is required to tackle deforestation and boost carbon capture and storage. In countries like Brazil and Indonesia, deforestation is leading to forests being a source of carbon rather than the carbon sinks that they should be. Illegal logging to make way for mining and agriculture are the main culprits. Brazil, DR Congo, Bolivia, Indonesia and Peru are witnessing the largest loss of primary tropical forest. The Grain for Green Programme in China is the longest running deforestation programme. The Nature Conservancy's Plant a Billion Trees Campaign also aims to reverse deforestation on a large scale in countries like China, Brazil and Colombia.

Sinal do Vale (Brazil), a retreat and bio-hub, is part of The Long Run network, Lata Foundation and Ecosystem Restoration Communities. It has made regeneration and restoring the remaining Atlantic Rainforest near Rio de Janeiro its core purpose through education for visitors and local communities.

Communities at heart

Capacity building for solid foundations

The global frameworks, technical know-how and technologies are available to address the challenges of transforming travel and tourism to a sustainable and resilient industry. Yet there is a high degree of fragmentation, and it is not easy to scale solutions. It takes time, but deadlines are looming. Social drivers, namely corporate responses and consumer consumption habits, not technical know-how, will be the cause of failure in achieving decarbonization by 2050, according to the Hamburg Climate Futures Institute.[29] This points to the need to change attitudes, mindsets and behaviours.

Capacity building in destinations and communities is critical to ensure the sharing of information, education and resources to help deliver the required outcomes. Capacity building encompasses a wide range of areas from policy, governance, skills development and education. This is where global institutions like the World Bank, UN Tourism, NGOs, non-profits, tourism boards, private sector work in partnership with local communities and DMOs to drive positive change. UNESCO runs extensive projects globally, regionally and by country funded by donor countries, such as China, Italy, the EU and Japan. The most significant focus areas for investment are education, culture, natural sciences, communication and information.

GSTC has an extensive selection of paid-for training courses from sustainable tourism, hospitality, business travel, DEI, accessibility and inclusion, through the lens of GSTC accreditation and metrics.

There are challenges where mutual trust is essential to guarantee that communities will be onboard. Often capacity building takes a top-down approach, applying frameworks that may not be so well suited to every location.

'Three challenges that need to be addressed are a new funding mechanism to help real sustainable tourism, increasing consumer awareness through education about the need to travel more sustainably, and finally joined up thinking between government and the private sector.'

Vicky Smith, Founder, Earth-Changers

Empowering local communities to take control of travel businesses and destination management is the most important way to drive resilience with the building blocks to survive future threats. Part of the upskilling process is ensuring that digital tools are democratized so that MSMEs can jumpstart operations in terms of bookings and customer engagement.

There are many ways to tap into skills training and resources across the travel industry. A popular way is to join a members' travel association. In Asia Pacific, Airbnb and PATA are partnering to uplift local communities through capacity building through digitalization.

The Caribbean Tourism Organization (CTO) considers capacity building as an integral function of their activities, working with member destinations and MSMEs at the forefront of climate change. In 2024, CTO rolled out a tourism resilience programme in partnership with The Travel Foundation, Global Tourism Resilience Centre and George Washington University.

With Caribbean destinations and communities being some of the most vulnerable to climate change and heavily reliant on tourism, adaptation and resilience are critical. Efforts such as the Inter-American Development Bank (IDB)'s ONE Caribbean Regional Programme will help support the necessary sustainable transformation through funding mechanisms, innovation and digitalization. CTO has provided resources and toolkits to support and drive resilience for community tourism businesses.

TIPS

- Public/private partnerships are fundamental to success to ensure that local communities reap the benefits of tourism, and gain the necessary skills through education and upskilling.
- Capacity building is a long-term ongoing commitment and should never be viewed as a one-off.
- Without an inclusive, collaborative and innovative approach to capacity building, efforts will be designed to fail.
- Resident surveys, online platforms and regular touchpoints are essential to keeping ahead of potential issues.
- Funding capacity building is a challenge but should be factored into budget planning.
- There are no better mentors than employees.

Community tourism for regenerative travel

Community tourism has transformed into a global phenomenon, empowering local residents, MSMEs and Indigenous People to lead and manage tourism businesses directly. Adopting a community-led model ensures that leakage is minimized, ensuring that tourism revenues remain locally, and are used to lift up education, resources and healthcare. The social aspect of community tourism is front and centre, but not to the detriment of environmental concerns. If working with aid organizations, NGOs or national tourism boards, communities can lean in for capacity building and access to resources, along with the necessary tools, digital or otherwise.

Enabling communities to lead and take ownership ensures many benefits beyond financial. Co-created travel and tourism products and services are designed with greater authenticity while meeting local needs and aligned with skills. This helps to elevate the visitor engagement and equally promoting and preserving local cultural and natural assets. For host communities, community tourism drives employment, skills development and agency for positive change.

Community tourism is particularly strong in the Caribbean in countries like Jamaica, with experiences steeped in the local culture showcasing food, music, culture and adventure. Community tourism works, providing home stays and getting to know the way of life of ethnic and indigenous peoples in rural and urban areas. The Long Run network reported that 54 per cent of its members co-create activities with local communities.[30]

The community-owned and -managed approach is not exclusive to emerging and developing countries. Scotland launched SCOTO – Scottish Community Tourism network, where there was a gap in the country's product offer. There is a consumer-facing brand BeLocal alongside the B2B members' network. Communities often feel overwhelmed by visitor demand with terms like invasion bandied around. The aim of sustainable development is to help communities to recalibrate tourism to put communities and environment first in regions like the Highlands and Islands. Including travel businesses and locals, creative ideas are brought to the table and aligned with the expectations for visitors looking for an authentic and impact-conscious experience. The target market is for visitors looking to live like a local resident. Initial public investment helped to launch the social enterprise.

REAL-WORLD EXAMPLE

DASTA building a community tourism network at scale

In Thailand, community tourism is an integral part of the country's sustainable tourism strategy. Organizations like Designated Areas for Sustainable Tourism

Administration (DASTA) – a public organization – have been working long and hard to develop a network of sustainable community tourism. Thailand is a highly popular destination with a diverse mix of visitors from Europe, China and intra-regionally. The Hollywood film *The Beach* brought global attention to Thailand's beaches with backpacking appeal, leading to overtourism of Maya Bay, which closed for a few years to enable it to recover. Since its reopening in 2022, daily visitors are limited. Yet overtourism is back on the agenda. Together with industry partners, the Thailand new Zero Carbon app for use by travellers, communities and tour operators is an important tool in promoting thousands of net zero routes, purchasing carbon credits and meeting net zero goals by 2027. A controversial tourist tax was shelved in 2024.

DASTA recognizes that each community is unique with its own needs such as economic, cultural and regenerative. Working with the GSTC ensures that global standards of sustainable tourism are met. Its communities are part of the Planeterra Global Community Tourism Network, tapping into their resources and funding.

DASTA helps to promote cultural/creative tourism or nature tourism, with a network of tourism clusters, collaborating with each community to agree on what works best in terms of its product offer, strategy and implementation. Working with the UNDP Accelerator Labs programme, DASTA enabled community tourism to co-create with people with disabilities to solve pain points and remove barriers for inclusion. A couple of designated areas like Nan in Thailand have been awarded Top 100 Green Destinations status.

TIPS

- Promoting community businesses is a no-brainer to help visitors know that their spending makes a positive difference locally.

- Travel businesses that measure and report the percentage amount of revenues left locally have a strong USP versus competitors.

- The creation of successful community tourism in rural areas helps to attract young people to stay and work locally, slowing depopulation.

- Taking a hyper-local inclusive approach to destination management will ensure that resilience is solid and future shocks can be dealt with quickly.

Indigenous tourism for inclusion and climate justice

Around the world, Indigenous Peoples use tourism as a vehicle for socio-economic sustainability by sharing their culture, language, traditions and stories after centuries of colonization and oppression to aid reconciliation. Powerful examples of indigenous tourism as a force for inclusive and sustainable growth are Canada, New Zealand, Australia, Latin America, Asia Pacific and the Caribbean. Indigenous People's connections with land and biodiversity make them essential actors in the fight against climate change. The track record for Indigenous Peoples is noteworthy for less damage has been inflicted on the environment, where deforestation rates are twice as low and emissions reductions are massively improved. With Indigenous Peoples responsible for 80 per cent of the world's diversity, their roles and voices are critical.[31]

In the same way that community tourism businesses are community-owned, indigenous travel businesses are owned by Indigenous Peoples. Indigenous Tourism Association of Canada is a global leader in advocacy, funding and marketing, where it launched the The Original Original accreditation programme to help visitors identify indigenously owned and managed travel businesses. It works in partnership with the communities and public sector with the aim to triple indigenous tourism revenue to CAD6 billion and create 60,000 jobs in nearly 2,700 businesses by 2030.[32]

In Vanuatu, there is an interesting intersection of regenerative agritourism led by indigenous communities leading to the creation of Regenerative Vanua, which provides certification aligned and certified by GSTC. Following a regenerative approach based on Indigenous knowledge and practices helps to protect the environment and drives climate resilience. This ensures the sustainable development of food tourism and regenerative agritourism experiences.

Aboriginal experiences are integral to countries like Australia, with its offer of guided tours, adventures and trips into the bush to gain insights into the world's oldest living culture. In Aotaeroa (New Zealand), Māori culture is integral to its identity and culture. The country's tourism strategy has been deeply influenced by indigenous principles. New Zealand's trailblazing move to regenerative tourism instigated a global shift in national tourism strategies.

TIPS

- Indigenous and regenerative go hand in hand, rejecting exploitation of the natural world.
- Tapping into Indigenous knowledge enables communities to be positive agents of change, driving inclusion and equality.

- Taking inspiration from diverse perspectives leads to creative, innovative and human-centric solutions.

- Indigenous Peoples should be further empowered and given a stronger voice and platform as they are at the frontline of the climate emergency.

- A multidisciplinary approach that combines scientific and local knowledge should prevail.

REAL-WORLD EXAMPLE

Planeterra puts social impact at the heart of its community activities

Planeterra is a social impact NGO founded in Canada and the non-profit arm of G Adventures, a global adventure travel business and social enterprise, created by Bruce Poon Tip. Planeterra supports and works with communities through a global network of partners and community tourism enterprise projects. A vital attribute of community tourism is that the businesses are owned and run by the community either as co-operatives, non-profit organizations or social enterprises. This approach ensures that tourism revenues are fairly distributed and benefit communities directly. For travellers, this provides a more rewarding, unique and authentic experience that equally empowers communities with a positive impact. Planeterra's 2030 goals are to improve the lives of 3.5 million people, generate US$1 billion in income for communities and enable 50 million visitors to experience local community-led experiences.[33]

Planeterra has formed new impact investor partnerships with Evaneos and Iberostar Foundation with a shared purpose to promote community centric businesses to reduce poverty and drive inclusion. The community tourism businesses are run by women, Indigenous Peoples, rural communities and youth.

Part of Planeterra's functions are the Global Community Tourism Network and the Global Community Tourism Fund, a financial lever to create positive change. The Global Community Tourism Network is free to join and enables MSMEs to engage with peers and receive online training and advice on capacity building. There are over 500 impact-driven businesses in the network across 80 countries. The Fund helps community tourism businesses to innovate and scale up their businesses, provided they meet sustainability criteria.

Planeterra also works with the IUCN to focus on protected areas in Vietnam and Peru, combining the Mediterranean Experience of Eco-Tourism (MEET) Network's nature-positive approach. Taking a joined-up approach ensures that both nature and communities thrive, while elevating the quality of the visitor experience.

Driven by purpose

Accreditation such as B Corp ensures transparency and accountability to strict ESG accounting standards. Adventure travel brands like Intrepid Travel and Exodus wear their B Corp badge with pride. The Travel Corp's not for profit arm, Treadright Foundation, is another best-in-class example of how to create positive impact for people and places by taking a purpose-driven approach. Initiatives include its net zero climate action plan and Make Travel Matter impact experiences for visitors. Exodus Adventure Travels Foundation takes a nature-positive position, working on conservation, regeneration, community building, disaster relief, equity and inclusion.

Impact travel businesses are increasingly making their mark. Journeys with Purpose won a Global Vision award from T + Leisure 2024 and is a member of the 1 per cent For the Planet network where 1 per cent of company sales are donated to environmental action. Hosted experiences are led by the park managers, NGOs, environmentalists and local communities with first-hand knowledge of the challenges that destinations are facing. This behind-the-scenes perspective is the definition of immersive nature-positive experiences, with a potentially life-changing impact, importantly creating a new army of climate activists. Partners include Tomkins Conservation, Global Forest Generation and One Tree Planted.

More and more businesses are driven by purpose, taking climate actions to do the right thing. Reportedly 66 per cent of Fortune 500 companies have climate targets in place, and of those they enjoy higher revenues (6 per cent more), higher profits (double) and emit substantially less (68 per cent less) on average than the 44 per cent that do not take action.[34] A common attribute for the climate committed is a chief sustainability officer to lead. Of those that signed up to climate action, many have joined the Science Based Targets Initiative (SBTi) to ensure that net zero is achieved by 2050.

If a business's sole purpose is financially driven then the travel industry will continue to be unsustainable, as decisions will be made in the headquarters of global companies and venture capitalists without consideration or input from local communities and destinations. Travel and tourism will continue to be considered to be a commodity that can be bought and sold.

'Where there are massive investments such as in resorts and cruise, when you follow the money trail, companies and investors want massive returns. Tourism becomes about the money, and not always about the people and places who live in those destinations. We must not lose sight of them.

We need to engage customers, businesses and all stakeholders of a destination so that everyone feels part of the regeneration of a place and creates the kind of tourism that spreads its benefits as widely as possible.'

Rochelle Turner, Global B Corp Impact Manager, Intrepid Travel

Emissions reduction entails different approaches including carbon offsetting, especially for scope 3 emissions that take place in the supply chain or downstream. Carbon offsetting involves compensating for emissions produced by purchasing carbon credits. However, with carbon offsetting scheme scandals and claims of greenwashing by international NGOs, offsetting should not be used to continue business as usual. Although the Science Based Targets initiative (SBTi) has announced that it does have a role to play.

TIPS

- Taking climate action is financially sound for business; purpose-led businesses will drive sustainable results.
- Carbon offsetting has been mired in controversy with some viewing it as greenwashing and should never be seen as a replacement for direct emissions reductions.
- Due diligence is required to ensure any carbon offsetting scheme is verified and as effective as it claims.
- Verification by international bodies like Gold Standard will help to avoid false schemes.

Education and awareness

Importance of education to be future-ready

Greta Thunberg has championed climate change awareness on a global scale. The climate strikes were especially effective in terms of the global mobilization of young generations through #FridaysForFuture strikes.

Generations Z and Alpha are more acutely aware of climate change, experiencing it first-hand as well as living through the fallouts from climate injustice that have been passed on from previous generations. Education is vital to ensure the necessary transition to meet climate targets and protect the natural world. Italy was the first country to make climate change education compulsory for school children. Globally, there is scope to expand teachings on the rights of the child to include the rights of nature.

Ecuador was the first county to legally give rights to nature. Mexico City updated its constitution to ensure nature is considered when passing laws. In Ireland, there is a campaign to consider the River Shannon as a legal entity, given equal rights with the right to thrive without human impacts such as pollution and extraction. The inspiration for this campaign is Whanganui River in New Zealand that was given rights due to action taken by the indigenous Māori community. There is a growing movement for the right for nature to thrive, championed by the Global Alliance for the Rights of Nature (GARN). The definition of nature is wide, covering rivers, mountains, plants and species. There are calls for ecocide to be considered a crime against humanity whether it impacts humans or nature.

With the escalating climate emergency, nature being called to have equal rights and the urgent need for behavioural change, education drives awareness and action at grassroots level. This starts with educating children from nursery onwards to change hearts and minds.

It is not just the young taking action. In Switzerland, a group of Baby Boomer women took on the government and won due to the country's poor climate policies that negatively impacted women's human rights due to greater risk of death from heatwaves. The European Court of Human Rights (ECHR) agreed that failure to meet 1.5°C targets is a violation of human rights. Climate justice cases will increase exponentially if there is a failure to act to limit global warming, led by ever more young people, angry at the world that they have inherited.

Building stronger links between education and industry

The global economy is transitioning to services, away from fossil fuels and exploitative industries. This opens up major opportunities for travel, tourism, transport, hospitality, mobility, food and drinks. Academia is increasingly connected to industry, through work placements for students, collaboration and research. With the need to accelerate decarbonization at scale, preparing the future workforce while upskilling the current workforce is challenging, notwithstanding the challenges posed by Gen AI and automation. Already institutions such as the University of Glasgow are taking an interdisciplinary approach offering a sustainable tourism and climate change course. The SunX Program is at the cutting edge of climate-positive travel education, offering the first ever climate resilient two-year diploma.

REAL-WORLD EXAMPLE

Edinburgh Napier University – lynchpin between education and the tourism industry in Scotland

Edinburgh Napier University (ENU) offers four undergraduate degrees in tourism, hospitality, management and festivals and events management, as well as postgraduate courses. Tourism courses sit in the business school, which includes a Tourism Research Centre. ENU is highly engaged with the travel industry in Scotland, through the Tourism and Hospitality Skills Group (THSG) part of Skills Development Scotland, across the private and public sectors.

Before budget cuts from the Scottish Government, three-day upskilling courses were offered for free, which enabled people working in tourism access to upskilling opportunities. Paid-for courses are still available including revenue management and destination development including resilience. This option to lean into education at different entry points in a career works well for travel employees and their employers.

Attracting and retaining talent remains challenging, with tourism students often working in retail rather than tourism due to more sociable and predictable hours during their degree. ENU provides a pipeline of future talent, putting undergraduates in touch with industry, and vice versa through collaboration, placements, networking and events. Working in partnership with industry, toolkits such as Tourism Hospitality Staff Induction help guide the transition to net zero, staff wellbeing, education and training, fair work, diversity and inclusion.

ENU is working already with other academic institutions like Heriot-Watt and Edinburgh University via TravelTech Scotland on understanding the wider implications of AI and robotics on travel and tourism. Taking an interdisciplinary approach enables efficiencies of scale and application. With AI simultaneously destroying jobs and creating new jobs, it is vital for educational institutions to keep one step ahead of disruptive forces.

'The way forward for tourism is taking an interdisciplinary approach such as linking tourism with health and social care in areas like AI, essentially taking a "tourism plus" approach.

We are heading in the right direction when it comes to tackling the climate emergency, but we need to overcome the challenges of how to scale up especially when the sector is so fragmented and the middle ground doesn't know how to engage with net zero targets.'

Anna Leask, Professor of Tourism Management, Edinburgh Napier University

TIPS

- Climate education needs to be fully integrated into the curriculum from early years.
- Travel, tourism and hospitality courses should consider incorporating in-destination placements whether related to tourism or adjacent industries to build a true connection with people and the planet.

Future workforce focused on positive impact

With the age of exponential growth and AI only at the beginning of its capabilities, new jobs in travel will emerge. There are already signs of what they might look like – travel data scientists, AI managers and, most importantly, travel climate scientists and impact managers. A combined approach will be required to ensure that travel businesses and their employees deliver a positive impact, with a strong understanding of climate science, STEM research, data, and destination management enabled by digital.

TIPS

- Sustainable travel, tourism and destination management courses need to adapt to incorporate the latest climate science.
- Each travel business and their employees can be empowered to be travel climate leaders through nature-based and conservation programmes.

Power of storytelling to change the narrative

Education and storytelling are key drivers of positive change in attitudes and behaviours to ensure that the preferred pathway of 1.5°C is achieved. The power of social media can amplify these important messages even further and reach every corner of the world in ways that previously did not exist. It is important for travel brands and destinations to work with influencers that are passionate about sustainability, positive impacts and biodiversity. Identifying influencers, with large or small followings, can bring a new layer of authenticity to brand engagement.

Platforms like Facebook, X (formerly known as Twitter), Instagram, WhatsApp, WeChat, TikTok, YouTube and Snapchat are among the leaders. Facebook and Instagram owned by Meta enjoy billions of monthly users where 64 per cent of consumers look there for travel inspiration.[35] Social media is where consumers spend a lot of their time and seek increased participation with brands and destinations. Consumers love looking for inspiration, communicating and sharing stories about their favourite experiences especially travel, food and lifestyles.

Word of mouth and user-generated content (UGC) have been the hallmarks of the web 2.0 era delivering strong returns where every US$1 invested in influencer marketing, US$6.50 is recouped in revenue.[36] Real people as content creators adds a layer of authenticity and emotional connection to interactions. Travel bloggers, vloggers, Instagram and TikTok influencers are increasingly used not just for inspiration and trip ideas but bookings. Some influencers like Alex Ojeda and JuliaGal enjoy millions of followers and exert much influence on where to go and what to do. Influencers bear a strong responsibility to equally promote positive travel behaviours.

Expedia even enables creators to sell travel products and services directly to their followers. Yet influencers do not hold as much sway as friends and family where 40 per cent of consumers look to their family and friends for travel inspiration.[37] The potential for social commerce has still a long way to go before it is fully realized.

For some of the most successful travel businesses, storytelling is foundational to their continued success. Tourism boards are investing heavily in the inspiration stage of the customer journey. VisitScotland is closing its information iCentres in popular destinations to focus on engaging with visitors at the inspiration stage of the customer journey.

TIPS

- The creator economy is booming and many travel brands enjoy high engagement from influencer marketing – due diligence is vital to ensure values are aligned between influencer and brand.
- The best-placed influencers are the closest to the business – locals, visitors and employees.
- Brands like Ryanair have taken a 'gonzo social media' approach engaging in an authentic and satirical way with big results but with high risks.
- To stand out in all the noise, authenticity and purpose will resonate and create emotional connections.

Human-centric travel business models

Google's planned removal of third-party cookies, stronger privacy laws like GDPR and web 3.0 are ushering in a new era where consumers are in the driving seat, in control of their own data and how businesses engage with them. More opportunities are opening for consumers to co-create together as well as exchange and trade their data for services that appeal to them. This moves the consumer from being passive to an active participant in engaging and co-creating with brands.

A new generation of travel start-ups are emerging putting human-centricity at the heart of their businesses, empowering individuals, employees and communities, and creating new revenue streams. Investors like Blackrock called out the importance of purpose for long-term profitability, while human-centricity focuses on human-centric design and provides solutions to real human problems and meeting their needs – from inclusion to climate change. Human-centric design understands that people, biodiversity and ecosystems are interconnected and demands a system-wide approach to solve complex problems with communities. Ultimately, human-centric business models enhance the triple bottom line and create shared value in an inclusive and fair way.

REAL-WORLD EXAMPLE

Bolder empowers communities with storytelling and shared value creation

Bolder works with the ATTA and destinations in Norway, Ireland and Rwanda. It is taking a bold leap forward in helping communities, individuals and small businesses to take back control and maximize their assets, marketing and customer engagement.

With access to Bolder's Universal Basic Digital Infrastructure (UBDI), people can co-create, share, collaborate and co-distribute their travel services, stories and experiences on their own terms. Stories can feature social content from trips, experiences and passions such as sport, hiking and food, for example. The UBDI is to democratize access to technology so that everyone can contribute and engage with autonomy. There is a shift in approach where technology is used to truly serve people, delivering value and driven by purpose.

The online platform and marketplace are born out of the adventure travel community, where sports, nature and adventure meet the creator economy. Founders Eirik Skjærseth and Anne-Sofie Engelschiøn are leading the next generation of human-centric, people-oriented travel businesses.

Bolder's strategy follows the research showing that the optimal tourism experience involves the highest levels of technology-enhanced co-creation. As opposed to industrial tools where the potential outcomes are predefined by a designer, the company is using convivial tools to build trusted personal connections. Convivial tools are tools that allow the user to impact the end results with their creativity, ultimately fostering a flourishing of innovation and new ways of collaboration. Moving away from centralized technology platforms and centralized thinking, the company aims to revolutionize travel so that it transforms away from its current extractive and industrialized model through a cooperative model.

The Bolder community enables destinations and local communities to earn while they create, hold events and earn revenue from activities. Partners include Rwanda Red Rock Cultural Rock and Eco Sails in Stavanger (Norway). There are two sides to the marketplace: the consumer co-op on one hand and the producer co-op on the other. On the producer side, customers are owned and shared. Working around key events and festivals, viral network effects are created. Bolder's vision of the future of travel is one where there is decentralization and a thriving creator economy at the grassroots community level, with benefits rippling system-wide.

'The traditional DMO model aimed to bring lots of people to a destination with no thought of resilience, sustainability or regeneration. Bolder is helping to build the next generation of DMOs from the bottom up through a trusted network economy. Here micro businesses driven by passion and purpose can set their own agenda and generate sustainable value and provide enriching visitor experiences.'

Eirik Skjærseth, Founder, Bolder

Summary

Resilience is the ability to deal with external shocks, adapt and bounce back stronger than before. Travel businesses and destinations continue to demonstrate unprecedented levels of resilience, constantly adapting to the triple planetary crisis of climate change, pollution and biodiversity loss. There are many organizations, especially non-profits, doing a great job in capacity building to ensure communities have the right funding mechanisms, resources and skills.

However, the traditional business model remains, with business as usual restored. Despite the global frameworks and pathways set out to avoid irreversible planetary tipping points, the actions of most governments, travel businesses and consumers do not go far enough. The 1.5°C pathway for sustainable development looks more like a 2.7°C disaster. Without true purpose that prioritizes positive and enduring impact, resilience strategies and action plans will ultimately fall short.

A new paradigm is required to achieve the global goals. This means putting aside short-term wins to focus on long-term benefits for all. The economic case for being climate-positive has been made. The technology, tools and resources are available. It is mindsets and behaviours that are stifling progress. A complete reset is needed to rebalance the community's needs and visitors' wants so that the relationship is mutually beneficial, decentralized and no longer extractive.

For resilient and sustainable transformation, a grassroots approach is required for destination management, adopting nature-positive solutions and empowering local people and Indigenous Peoples. A holistic approach that elevates local voices and engagement will ensure that everyone can play their vital role in securing a liveable future.

Technology for good is a key enabler in achieving not just sustainable transformation but creating an army of impact-conscious travel businesses, big and small. With the inevitable rises in temperatures despite adaptation and mitigation strategies and the carbon budget due for depletion by 2029, the time for action is now.

Endnotes

1 WTTC (2024) wttc.org/news-article/travel-and-tourism-set-to-break-all-records-in-2024-reveals-wttc (archived at https://perma.cc/EGL8-359U)

2 World Economic Forum (2024) www.mckinsey.com/~/media/mckinsey/business%20functions/risk/our%20insights/building%20a%20resilient%20tomorrow%20concrete%20actions%20for%20global%20leaders/wef_building_a_resilient_tomorrow_2024.pdf?shouldIndex=false (archived at https://perma.cc/MXG6-ZEB6)

3 University Notre Dame, Notre Dame Global Adaptation Initiative (2024) gain.nd.edu/our-work/country-index/rankings/ (archived at https://perma.cc/ZJE7-TQV8)

4 World Bank (2024) climateknowledgeportal.worldbank.org/country/maldives (archived at https://perma.cc/9DH3-5EYY)

5 World Bank (2022) www.worldbank.org/en/news/press-release/2022/06/02/world-bank-supports-maldives-to-leverage-digital-technologies-for-development-and-climate-resilience (archived at https://perma.cc/JZK8-ATEX)

6 Benjamin H Strauss et al., 2021 Environ. Res. Lett. 16 114015 iopscience.iop.org/article/10.1088/1748-9326/ac2e6b (archived at https://perma.cc/YN9Q-DC9M)

7 Iberostar Wave of Change (2024) waveofchange.com/resource/wave-of-change-2023-year-in-review/ (archived at https://perma.cc/HD6J-NGPU)

8 Convention on Biodiversity (2023) www.cbd.int/gbf/goals (archived at https://perma.cc/524T-T5Y5)

9 WWF (2023) www.wwf.org.uk/sites/default/files/2023-05/WWF-Living-Planet-Report-2022.pdf (archived at https://perma.cc/8AQM-KTXW)

10 Ibid

11 United Nations (2022) www.un.org/en/climatechange/science/climate-issues/biodiversity (archived at https://perma.cc/NXZ5-GG4R)

12 Ibid

13 Jane Goodall Institute (2024) janegoodall.org/ (archived at https://perma.cc/B7K5-T74N)

14 Animondial/WTTC (2024) research.wttc.org/nature-positive-travel-and-tourism (archived at https://perma.cc/T38H-HP3B)

15 University of York (2024) www.york.ac.uk/news-and-events/news/2024/research/climate-change-biodiversity/www.science.org/doi/10.1126/science.adn3441 (archived at https://perma.cc/NQ85-DJBP)

16 NOAA (2019) www.noaa.gov (archived at https://perma.cc/9CUF-JL5H)

17 IUCN (2024) www.iucnredlist.org/resources/summary-statistics (archived at https://perma.cc/ZXB6-P2L6)

18 Rewilding Chile (2024) www.rewildingchile.org/en/ (archived at https://perma.cc/W3MT-3GHK)

19 IMF (2020) www.imf.org/ (archived at https://perma.cc/4ZW3-QGV2)

20 The Nature Conservancy (2024) www.nature.org/en-us/what-we-do/our-priorities/ (archived at https://perma.cc/P3LB-5XR4)

21 African Development Bank (2024) www.afdb.org/en/knowledge/publications/
tracking-africa%E2%80%99s-progress-in-figures/human-development
(archived at https://perma.cc/BK4B-7VTD)

22 Greenview (2022) greenview.sg/wp-content/uploads/2022/12/Green_Lodging_
Trends_Report_2022.pdf (archived at https://perma.cc/6DJN-45PX)

23 Global Himalayan Expedition (2024) www.ghe.co.in/ (archived at https://
perma.cc/B8U4-XBVH)

24 Schiphol Group (2024) www.schiphol.nl/en/schiphol-group/page/economical-
energy/ (archived at https://perma.cc/9PN8-HN98)

25 Greenview (2022) Green Lodging Hotel Report

26 IRENA© (2024) www.irena.org/Data/View-data-by-topic/Capacity-and-
Generation/Country-Rankings (archived at https://perma.cc/7HV4-MM9L)

27 WWF (2024) www.worldwildlife.org/stories/what-do-sea-turtles-eat-unfortunately-
plastic-bags (archived at https://perma.cc/9PN8-HN98)

28 Greenview (2022) greenview.sg/wp-content/uploads/2022/12/Green_Lodging_
Trends_Report_2022.pdf (archived at https://perma.cc/77MA-CCZJ)

29 Anita Engels, Jochem Marotzke, Eduardo Gonçalves Gresse, Andrés López-
Rivera, Anna Pagnone, Jan Wilkens (2023) (eds.) Hamburg Climate Futures
Outlook

30 The Long Run (2024) www.thelongrun.org/wp-content/uploads/2024/05/2023-
Impact-Report_Spread.pdf (archived at https://perma.cc/33X9-97AB)

31 Convention on Biological Diversity (2022) www.cbd.int/kb/record/newsHeadlines/
135368?FreeText=protected%20areas (archived at https://perma.cc/VPM5-JUBC)

32 Indigenous Tourism Association of Canada (2024) indigenoustourism.ca/
wp-content/uploads/2024/03/ITAC-2024-25-Action-Plan-Final.pdf (archived at
https://perma.cc/R3ZY-UGYZ)

33 Planeterra (2024) planeterra.org/impact/ (archived at https://perma.cc/4JZ9-
PWH6)

34 Climate Impact Partners (2024) info.climateimpact.com/hubfs/
ClimateImpactPartners_FortuneGlobal500_2023_FINAL.pdf (archived at
https://perma.cc/P79Z-PKMV)

35 Booking.com (2023) partnerships.booking.com/resources/article/travellers-
reveal-their-top-sources-trip-inspiration (archived at https://perma.cc/
HZH6-S4TB)

36 Forbes (2024) www.forbes.com/sites/forbesagencycouncil/2024/01/05/
the-power-of-influencer-marketing-your-strategic-investment-for-success/
(archived at https://perma.cc/GG24-VP8N)

37 Booking.com (2024) partnerships.booking.com/resources/article/travellers-
reveal-their-top-sources-trip-inspiration (archived at https://perma.cc/
N8QQ-KWKJ)

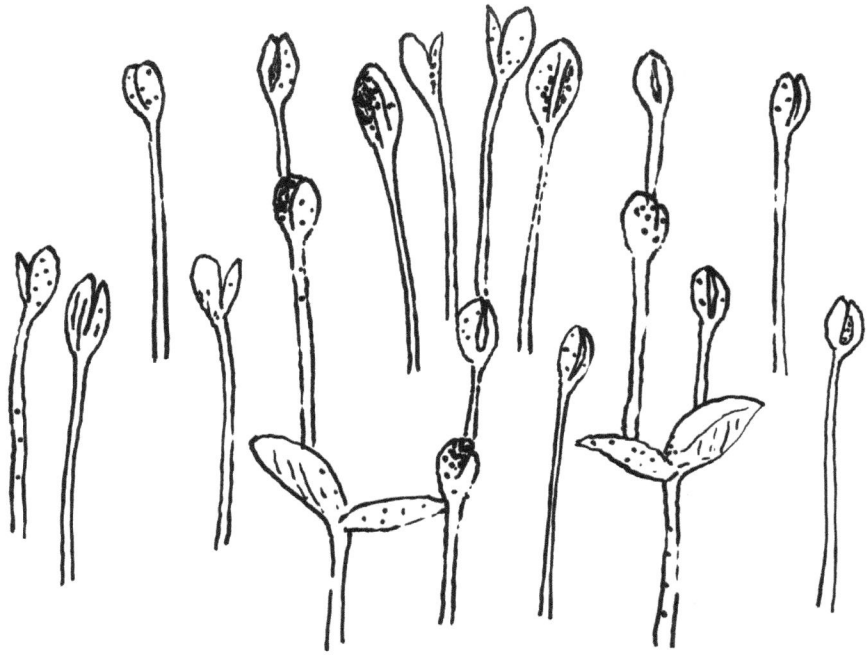

07

Pathways to accelerate sustainable transformation

Introduction

The need for sustainable transformation is ever more pressing as the milestone of 2030 approaches and focus then shifts to reach net zero emissions by 2050. Signs are already visible that the world will overshoot the optimum 1.5°C pathway as collective commitments are not bold enough. Resistance to change carbon-intensive activities and behaviours persists. Consumers show signs of fatigue when it comes to making the most sustainable choices or actions. They are confused by the mixed messages they receive from government and increasingly look to travel businesses for guidance in how to travel more sustainably. The 'say-do' gap between intention and behaviour remains a major barrier to travelling with minimum impact. Businesses turn to governments to instigate the necessary levers to accelerate positive change through regulation, funding, taxation and policy.

Without urgent action from all stakeholders to transform, a vicious circle will lead to higher temperatures through positive feedback loops. More extreme climatic events and the devastating consequences of breaching tipping points will be unleashed ever more frequently. The natural world and humanity will pay a heavy, and often deadly, price if there is failure to act urgently.

The risks are high where travel and tourism are not managed in a sustainable way such as overtourism and inequalities are allowed to run rampant. This imbalance leads to anti-tourism sentiment where local authorities erect barriers through taxation and stricter capacity limits. With the tide turning on excessive growth, the degrowth model may garner interest from tourism boards and communities, looking for an alternative pathway. The degrowth model may well work for some destinations but not all, especially emerging countries at the beginning of their socio-economic development cycle.

Living in an exponential age where technology advances in leaps and bounds exacerbates the challenges. Technology by nature causes disruption. Gen AI and automation may distract from the task of preserving planetary health as job losses mount. Ever-increasing levels of hyperconnectivity and personalization raise individuals' expectations that may clash with the collective good.

The pathways to accelerating resilient and sustainable transformation rest on harnessing the power of technology to deliver the best outcomes for people, places and planet to thrive and flourish. Of all industries, travel and tourism are in the most unique position to inspire and lead from the ground up within the global framework of what success looks like. Embedded in the fabric of local communities, travel businesses have an important role to speak up and empower their communities, employees and visitors to be champions of positive impacts. Travel with purpose justifies a pro-tourism approach that is inclusive, integrated and never acts in silo.

The writing is on the wall for unsustainable business practices with mandatory ESG reporting introduced in the EU through the Corporate Sustainability Reporting Directive (CSRD). With larger companies taking more accountability for their actions, the onus is on smaller companies to match these to stay relevant and competitive.

The paradigm shift to a value-driven model that is planet-friendly is already taking place, where economic growth is decoupling from climate and biodiversity targets. There are pockets of good work being done, but it is not joined up or holistic. The challenge is to scale and speed up the transformation required to succeed and create a future fit for all.

Technology for good

When it comes to technology, travel has been at the forefront of advances for decades. Some of the largest companies in travel are digital native like Trip.com, Expedia, Booking Holdings and Airbnb, if not AI native like Google. Technology underpins everything where data is power. Yet digitalization of travel services has equally contributed to the challenges faced especially in Europe today with low-cost carriers democratizing city breaks and Airbnb disintermediating lodging. Overtourism is a major hangover of democratized and cheap travel.

Technology for technology's sake is dangerous and leads to a zero-sum game. Instead, it shines when used for problem-solving, driving inclusion, and acts as an enabler of the global goals. There are thousands of smart cities all over the world, trialling and testing new technologies to deliver the net zero agenda.

'Technology should be seen as a means to an end not an end in itself. Technology was touted as the solution for over-crowding, yet we've not seen anything that is truly scalable to solve the challenges of over-crowding in destinations. The way to solve such challenges is for tourism to be integrated into the heart of urban planning from the get-go.'

Christopher Imbsen, VP Research and Sustainability, World Travel and Tourism Council

Smart cities at the forefront of sustainable transformation

Cities are increasingly a magnet for people to live and work with urbanization a core driver of continuing economic growth into the next century. Yet this leads to cities being a key contributor to carbon emissions, putting cities and their communities at the forefront of the fight against climate change.

In Europe, some cities are moving faster than others. The NetZero Cities Mission encompasses 112 cities across 35 countries that aim to be carbon neutral by 2030, an impressive 20 years earlier than the net zero milestone of 2050.[1] Iconic destinations include Barcelona, Amsterdam, Copenhagen, Dublin, Paris and Rome. The programme is staggered with tiers of cities entering two years of accelerated decarbonization, with lessons learnt and shared with future city participants.

Key areas to tackle for climate neutrality include smart mobility, transport, data and digitalization as well as social innovation and governance thanks to technology such as AI and IoT. The EU is funding the initiative with a systematic and iterative approach to accelerated transformation for decarbonization where technical knowledge is not enough and systemic transformation is required across all areas of society.[2]

REAL-WORLD EXAMPLE
Copenhagen is a beacon for smart city innovations powered by data and citizen engagement

Copenhagen continues to lead the pack when it comes to smart cities – aiming for carbon neutrality in 2025. A critical pillar of its success is its test and learn approach, using data and citizen engagement to test new digital urban solutions through its Copenhagen Solutions Lab. Projects include local measurement of air pollution and the use of AI and data to improve energy efficiency for optimizing energy use. The localized approach using the city as the testing ground enables hyper-local testing to then scale up across the city, or even nationally and beyond. Furthermore, the quadruple helix approach for innovation is used to ensure that citizens are involved in the innovation process along with academia, private and public sectors.

Copenhagen's 2025 climate plan focuses on four key areas: energy consumption, production, reduced emissions mobility and city administration. The 2035 plan aims to achieve climate positivity by 2035 in terms of direct carbon emissions and reduce emissions by residents by half compared to 2019 levels amounting to a reduction of 5 CO_2 tonnes per capita.[3]

Travel is highlighted as an area for targeting along with buildings, infrastructure, consumption of food, clothing and electronics. From embracing urban nature planning, the circular economy and lower emissions transport such as bicycles, the city is on a mission to lead the world in the net zero transition.

A central tenet to Copenhagen Solutions Lab to achieve sustainability is embracing the circular economy through its Circular Copenhagen initiative. This encompasses moving away from an extractive and waste of finite resources approach to one of reduce, reuse and recycle. Even in frontrunning climate-conscious cities like Copenhagen, there is a lack of progress in adopting circularity, with Denmark's economy only 4 per cent circular compared to the global average of 7.2 per cent.[4]

The Copenhagen Lessons for the global architecture sector are a shining light in how to develop the built environment that leaves no one behind with 10 principles recommended to follow regarding architecture and construction. The goal is to leave no waste or carbon footprint.

Copenhagen is part of the Carbon Neutral Cities Alliance (CNCA) along with cities like Helsinki, Rio de Janeiro, San Francisco, Toronto, Glasgow and London that aim to speed up deep carbon reduction. It is on track to achieve 100 per cent reduction of emissions over 2005–2025 where energy accounts for 66 per cent and transport 34 per cent of the city's emissions.[5] The city took part in the AI4Cities EU initiative that saw pilots of different data- and tech-driven solutions. Examples include Holoni using AI and blockchain to boost solar energy use and SPIKE using IoT, AI and the cloud for energy efficiencies. Tapping into prosumers (consumers who participate in the design and creation of innovative solutions), enables Copenhagen to make strides in achieving its goals to grow economically and demographically while achieving carbon neutrality.

TIPS

- Boosting climate resilience through scalable decarbonization by testing at the local level to then scale up is essential for destinations to ensure they are future-proofed and on track for net zero.
- Embracing circularity is key to sustainable transformation, where AI is already used in areas such as food waste in hotels and recycling.
- Prosumers are the key to advancing progress in carbon reduction through their proactive and participatory engagement in climate and nature-based solutions.
- The road to net zero should not be a zero-sum game with losers and winners but ensure that residents and visitors are part of the energy and mobility transition and reap the benefits.
- With the global circularity gap widening due to lack of progress, urgent action is required to make a circular transition.

Not every destination or travel business can afford to create a supercomputer or integrate sophisticated AI and IoT networks. There are other less costly ways to incorporate AI to improve services that are helpful for visitors and time-saving for travel businesses, especially those aligned with sustainable and regenerative practices.

Balkan Natural Adventure that operates tours on the Peaks of the Balkans has launched an AI chatbot on its website, enabled by ChatGPT. This provides useful tips on hiking and trekking along with general travel information. Younger generations are much more comfortable using AI-enabled customer service chatbots and functionality than older generations so this move by an adventure travel company helps it to keep ahead of customer expectations and stay relevant.

REAL-WORLD EXAMPLE
Barcelona – flashpoint for anti-tourism sentiment due to overtourism and inequalities

Barcelona (Spain) has a long history of problems with overtourism where its global city destination success has led to excessive tourism demand which has in turn led to tensions with the local community. It was one of the first destinations to trial big data and IoT, sourcing from telecommunications data and sensors to help manage visitor flows. In 2023, it received 12 million visitors (domestic and international), dwarfing the local population.[6] In July 2024, crowds of residents took to the streets to protest with some unpleasant exchanges with visitors. The city will need to work harder to illustrate to its residents the value and benefits that tourism brings to redress the sense of injustice felt locally.

As part of its long-term commitment to transform to responsible and sustainable tourism, digital technology is one of the tools used to address the challenges faced. Key strategic priorities for the Catalunya Tourism Board include decarbonization, biodiversity protection, diversity, inclusion, visitor dispersal and innovation to future-proof the wider region.

To promote collaboration and dialogue with the community, the local government created an open-source digital platform for residents to share their concerns and perspectives. Decidim.Barcelona gives citizens a voice and the opportunity to participate and shape city life through citizen-led initiatives. Barcelona prides itself on being a digital city, promoting inclusive access to digital technology for everyone of all ages. There is a local Tech Tourism Cluster that works to drive innovation across the entire travel supply chain supported by the Catalunya local government.

The creation of a local digital twin of Barcelona that encapsulates traffic, people flows along with multiple other indicators in real-time helps in terms of urban management and planning. The concept is based on the 15-minute model for cities according to Moreno, where services should be within easy reach of residents.[7]

Built in partnership with the Barcelona City Council, Barcelona Supercomputing Centre and University of Bologna (Italy), the platform will move from a static visualization of the city to a scenario-planning tool with the use of quantum computing, AI and simulations to determine outcomes. The use of real-time data from multiple sources makes the digital twin a powerful planning tool and tests the viability of areas such as electric vehicle charging stations, for example. The plan is to replicate the digital twin 15-minute model city, starting with Bologna. The digital twin is an important tool in future urban planning. However, the issue with short-term rentals and lack of access to affordable housing for residents, means that such technologies will not solve the root problem of anti-tourism.

Barcelona's open-source and innovative approach to digital boosts transparency, accountability and ensures a more equal playing field for local communities in how their city is managed. The city is home to the Global Observatory of Urban Artificial Intelligence, in partnership with Amsterdam and London. The city has an ethical AI strategy and several AI initiatives including AI beach capacity monitoring using thermal cameras with privacy by design. For example, the Mercè project used AI to identify what makes a city liveable based on residents' opinions.

TIPS

- Smart cities demonstrate what it is possible to achieve in terms of climate-positive city living when all stakeholders work together on testing innovative solutions.

- Using cities as the backdrop for living labs with a test and learn approach empowers start-ups and communities to contribute and be engaged with change.

- Not every project will work. Part of building resilience is accepting and adapting to failure, bouncing back quicker every time.

- Funding is a critical lever – creative industries must not be left out of climate transformation.

- Providing an open platform for collaboration with communities helps to ensure residents feel they are equal participants in tourism planning decisions.

- AI requires an ethical and inclusive strategy, and requires capacity building in communities and upskilling of employees.

- Digital twins are where ideas can be tried and tested in different scenarios.

- The use of spatial data in geographical digital twins is increasingly important in driving adaptation to climate change.

Climate scenarios and forecasting powered by geospatial data

Climate data is playing an ever-greater role in destination risks and resilience. Travel and tourism scenario planning is not an exact science, at times caught off guard by black swan events and unknowns. Some threats and challenges are easier to predict than others. Long-term drivers are easier such as demographics, generational shifts, migration, falling fertility rates, urbanization, climate change, food security and energy security. Wildcards like geopolitics, wars, cultural revolutions, breakthrough technologies, innovation and natural disasters are less so.

Success can only begin with a baseline measurement for targets to be set, with ongoing measurement to check progress. Data is already everywhere and will continue to explode when the new era of hyperconnectivity arrives. With so much data, it can be overwhelming to see what is essential.

The next generation of Gen AI climate technologies points to ways that big data and AI will help with adapting to ever more extreme weather events. NVIDIA, one of the largest companies in the world, launched its Earth-2 digital twin cloud platform where simulations help with adaptation planning. Building long-term tools like Climate Central's Coastal IDEM digital elevation model is important for sustainable transformation and resilience, leveraging satellite light detection and ranging (lidar) data, AI neural networks and 3D rendering along with Google Earth. Such tools help in terms of planning for risks such as rising sea levels and coastal flooding.

In Europe, the Copernicus Climate Change System provides highly useful weather forecasts to 2099 to help with future climate scenario planning. The metrics include temperature, fire danger and precipitation forecasts. Such forecast indicators like snow indices have helped winter tourism resorts in Europe to adapt to a warming climate. Already climate scenarios for tourism demand in Europe have been developed based on 1.5°C, 2°C, 3°C and 4°C of global warming. The Joint Research Centre's findings show how southern European destinations will be the most affected on the coast, forecast to see a decline of 10 per cent in tourism demand in the peak summer season by 2100, compared to a 5 per cent increase for Northern and Central Europe.[8]

The EU is taking its climate adaptability one step further, with the launch of Destination Earth, called DestinE, powered by Finland's supercomputer and AI to provide a digital twin of Earth to adapt to natural disasters, climate change and create the relevant enabling policies.

Early warning systems and the next generation of satellites will provide more accurate earth observation (EO) data to inform global warming scenarios. The results will help destinations like the Mediterranean to be more prepared for risks like forest fires if integrated into planning tools.

Tech giants like Google provide open-source climate data from Google Earth Engine, driven by the cloud and AI to help with infrastructure and resources planning. This approach enables public and private sector players to gain access to 50 petabytes of geospatial data to help with resilience and risk planning. Google partnered with Climate Engine to form SpatiaFi to combine geospatial and economic data to help infrastructure, transport, finance and public sector businesses. The platform leverages data past and forecast and real-time data from remote sensors to build financial resilience. In the case of CN Rail (Canada), SpatiaFi data was used to manage wildfire risks in real-time, protect trains and infrastructure, and for forecast planning.

TIPS

- Real-time geospatial data such as climate and extreme events with predictive analytics will increasingly become the norm for resilience and scenario planning, rather than the exception.

- Real-time geolocation mobile data from smartphones has seen much interest from DMOs, mobility players and travel businesses while data privacy is a major factor to consider.

- With ever-increasing reams of data, having a robust data strategy in place along with upskilling employees in areas such as data science and climate science will become more standard.

'Ways to drive sustainable transformation include fostering collaboration, innovation and diversification, empowering local communities and adopting a tech-driven approach to measuring success.'

Jeremy Sampson, Chief Executive Officer, The Travel Foundation

---

REAL-WORLD EXAMPLE

Belize takes a data-driven approach for sustainable tourism development and destination management

Destinations like Belize are using geographical information system (GIS) technology as provided by Esri to understand the impacts of climate change and develop their long-term sustainable tourism development using location data. Belize like the rest of the Caribbean is at the forefront of climate change, with storms, hurricanes and flooding, exacerbated by its low-lying position making it vulnerable to sea level rising.

GIS has been instrumental to the National Sustainable Tourism Master Plan to 2030. Working with different climate change scenarios, the GIS data has informed climate action plans for select destinations to ensure they can adapt and thrive, looking at scenarios in 2050 and 2100. Taking a spatial data-driven approach to land and tourism planning helps to direct investment in a more targeted way. It has led to a refresh in how Belize is approaching planning by adopting the National Strategic Tourism Spatial Framework, with a holistic view of the entire country encompassing hubs, nodes and corridors.

The country is tackling climate change, conservation and tourism together, taking a proactive managed approach to future demand. It is thinking system-wide in managing, protecting and adapting its urban and coastal areas to be resilient and thrive where tourism accounts for 50 per cent of GDP.[9] Part of its 80 targeted actions to 2030 include limits on visitors and boat traffic where adventure travel and hiking are regarded as a value segment to target. The plan aims to showcase living culture along with a more joined-up approach to Maya sites.

Belize Barrier Reef Reserve System, the world's second largest reef, is a UNESCO-protected site. The government of Belize is using the latest science and technology to ensure that biodiversity, ecosystems and communities are protected against the impacts of climate change. Belize was part of the UNESCO eDNA programme to check marine biodiversity health across select sites. Belize partners with Microsoft to use its AI cloud platform, Azure, for the AI for Belize National Marine Habitat Map programme to collect better satellite and spatial data.

Destinations and travel businesses continue to work with anonymized and GDPR compliant geolocation data to understand visitor flows in a more granular way to inform destination management. Exodus partners with NatureMetrics for next-generation monitoring of biodiversity through environmental DNA (eDNA) and machine learning to assess its nature impacts and risks. This proactive approach also aligns Exodus with global science-based standards on biodiversity and the recommendations of the Taskforce on Nature-related Financial Disclosures (TNFD). Exodus's trips and customers can get involved in collecting eDNA from wetlands during their trips and contribute to the eBioAtlas initiative through citizen science.

> ### TIPS
>
> - Real-time geospatial data is useful for managing visitor flows and for predictive analytics, however, data privacy and security must be protected.
>
> - The human right to freedom of movement must be upheld, however, where travel is deemed a want rather than a need, impacts should be addressed.
>
> - Forecast models should look beyond traditional business and leisure tourism demand to incorporate migratory shifts due to climate change.
>
> - International student flows are an untapped source of tourism demand, often excluded from official tourism statistics but critical for building brand ambassadors and the visiting friends and relatives (VFR) market.

Levers for change

Sustainable transformation in travel requires system-wide change across economies and societies to adapt to the new climate realities of a warming planet and population growth. By 2060, it is forecast that there will be a population of 10 billion people worldwide.[10] By 2050, 70 per cent of the population will be urban.[11] There are many levers that governments and local authorities can use such as policy, taxation and investment to ensure sustainable development and an enabling environment for the right type of value creation. The World Economic Forum's Travel and Tourism Development Index ranks Türkiye, Malaysia, the UAE, Indonesia and Cambodia as having the best policy and enabling conditions for travel and tourism in descending order.[12]

Environmental taxes and rising prices

The race to decarbonize transport and, in particular, aviation is heating up. Some airlines and countries are going it alone, contrary to IATA and ICAO that advocate the need to work within the airline framework of the Carbon Offset and Reduction Scheme for International Aviation (CORSIA). The airline industry advises against applying additional environmental taxes and surcharges as it states that there is no proof that taxation has a positive effect on CO_2 reductions. Faced with the scale of transformation required and the costs involved, airlines are breaking with the official industry line.

In the EU, the target is for 1 per cent of fuel to be SAF by 2025, 6 per cent by 2030 and 70 per cent by 2050.[13] This is part of the region's Fit for 55 initiative which requires every flight departing from the EU to include SAF. Airlines like Air France are the largest consumers of SAF and raised prices in 2022 via a surcharge to fund the transition to SAF.

It is not just a case of price hikes that consumers face but additional surcharges. Lufthansa's Environmental Cost Surcharge is a sign of further taxes to come. The airline will charge from EUR1 up to EUR72 depending on final destination from 2025. Lufthansa aims to reach net zero by 2050 and that requires ramping up SAF, paying for changes in the European Trading System and CORSIA. The costs of sustainable transformation will ultimately be passed onto consumers.

In Singapore, the Civil Aviation Authority of Singapore announced the introduction of a green fuel levy from 1 January 2026 to fund the transition to SAF, going from 1 per cent to 3 per cent to 5 per cent by 2030. The SAF mandate requires that all departure flights from Singapore are blended with SAF for a drop-in solution from 2026. The transition away from fossil fuels will be covered in part by rising flight prices, passed onto the consumer. The air passenger levy price will vary based on flight distance from US$3, USD$4 to US$12 for short, mid and long haul respectively.[14]

Singapore is already well placed to ramp up SAF where Neste, the major global producer of sustainable and renewable fuels, has doubled its production capacity at Changi Airport. However, the challenge is scaling up production regionally across Asia Pacific. Singapore is looking at additional areas such as electrification of vehicles, renewable fuels and more efficient air traffic management. Aviation is a central pillar of Singapore's competitiveness. The creation of the International Centre for Aviation Innovation will provide a testing ground for ideas, concept testing and simulations, thanks to a new digital twin. Transforming the workforce is fundamental to success where upskilling of existing staff for sustainable aviation is required, as well as ensuring the future workforce are sustainability ready. This requires working closely with higher education and even earlier to reshape curriculums and attitudes.

Tourist taxes are increasingly being adopted by destinations struggling to keep up with the investment required to manage tourism in a sustainable way. In Scotland, in May 2024, the Visitor Levy (Scotland) Bill was passed that allows individual local councils the flexibility to apply a visitor levy from 2026 after consultation with businesses and communities. The levy will amount to 1 per cent of the cost of an overnight accommodation stay.

Destinations from Bali, Barcelona and Venice to New Zealand are using taxation as a key lever to tackle overtourism; however, it can only be effective when integrated into a holistic approach to sustainable tourism that puts communities first.

TIPS

- Any form of increased pricing is difficult for travel businesses to justify to their customers; however, where there are overtourism pressures and funding gaps, it is a necessary step.

- Taxation is not a panacea and should be reviewed at regular intervals to see when it can be withdrawn or at least lowered.

- Providing greater visibility on how and where taxation is used locally is vital for both visitors and residents for transparency on how money is spent and benefits accrued.

- Visual branding helps to showcase which local businesses benefit from sustainability-led taxation.

Rewarding and incentivizing sustainable behaviours

Behavioural science and nudge techniques are often used to encourage the uptake of more sustainable behaviours through rewards, incentives and gamifying the experience. Such nudges in the right direction are increasingly introduced into airline loyalty programmes.

'The ongoing education of travellers travelling well, with a light but positively impactful touch is critical to the preservation of the values and value of travelling. There has never been a more important time to be more informed as a traveller, aware of one's own carbon footprint, the impact of your travelling and being prepared to mitigate it. However, it's also about recognizing the positive contribution you can make, and how to add value and make a difference to communities' lives. It's about being a more conscious, informed and responsible traveller.'

Fiona Jeffery, Founder and Chair, Just a Drop, and Chair, Institute of Travel and Tourism Sustainability Committee

Qantas has integrated green rewards into its frequent flyer loyalty programme with its Green Tier. Rewards and bonus points can be earnt for sustainable purchases and lifestyle choices across six different criteria. Options include carbon offsetting the flight, choosing sustainably accredited hotels and wine, installing green energy or offsetting at home. Charity donations include supporting the Great Barrier Reef Foundation.

Business air travel has a larger carbon footprint than leisure travel. Etihad launched its Corporate Conscious Choices to help its corporate clients to meet their ESG targets and reward sustainable behaviours. Options include bulk buying SAF to use to offset scope 3 emissions or choosing the green surcharge to fund offsetting projects such as reforestation or community projects.

Members of Flying Blue, Air France-KLM's loyalty programme, were the first to be able to buy SAF with their miles/points in return for earning status since 2022. Lufthansa launched its green fares programme in 2023, offering consumers pricing options with sustainability measures built in such as offsetting and funding SAF.

However, greenwashing continues to surround European airlines' sustainability claims regarding offsetting schemes and use of SAF. In April 2024, the European Commission (EC) and national consumer agencies started an investigation into 20 airlines and their environmental claims. The EC wants to ensure that claims are accurate and scientifically backed up.

There is an ongoing debate about where frequent flyer programmes should be banned as they reward and encourage more flights and associated emissions. Wizz Air's launch of a flight subscription model in 2024 sits counter to the need for less if not flying at all.

TIPS

- The path to decarbonize aviation will not be easy – for some, it will not be bold enough while for others it will be too extreme.

- Already there are countries and airlines breaking away from CORSIA; it could be time for a review of where and how aviation fits in nationally determined contributions.

- Green air taxes will become ever more prevalent, yet they must be transparent and not punitive to ensure a fair transition.

- Gamifying and incentivizing uptake of sustainable options would work more effectively than enforcement.

- Greater transparency is required so that consumers can fully understand how their financial contributions are being used.

- While governments and airlines are at loggerheads, consumers increasingly feel confused about what information to trust and how to make sustainable travel choices.

Adopting a start-up mindset through innovation and agility

For MSMEs the ability to influence system-wide change is not easy; therefore, travel businesses must be ever more creative, adaptive and agile, driven by purpose. By taking a leaf out of the start-up mantra, travel businesses can problem-solve, run pilots, test proof of concepts, learn from successes and failures. Most importantly, the ability to bounce back quickly and move on even quicker is paramount. Travel tech communities are already working in partnership with the travel industry in Dubai, Barcelona and Scotland, and this collaborative approach helps to forge public/private partnerships where a fresher approach helps leave legacy frameworks and ways of working behind. Successful start-ups build a minimum viable product (MVP) that they take out quickly to market, engage with early adopters, learn from feedback and iterate rapidly.

With large companies such as hospitality players setting annual budgets, even before time, MSMEs have a unique competitive advantage to innovate and scale quickly especially if part of a member network. Not every brand will become the next tech unicorn, but they can make a positive difference.

New travel start-ups making waves tend to focus on mobility services, operational efficiencies or removing pain points across the traveller journey to make it seamless and personalized. Travelperk and Mews saw their valuations move into the US$1 billion unicorn territory with fresh funding rounds. Other travel start-ups enjoying success include Klook, Get Your Guide and Guesty, the latter for short-term rental property management. Brands like Fora Travel are democratizing travel by enabling anyone to sell travel as advisers on their platform if they have the right skills and love of travel.

Others focus on using tech for good such as Pachama in the fight against climate change. Pachama deploys AI and geospatial or earth observation data to help Fortune 500 businesses like Airbnb to invest in nature through reforestation and verified carbon projects. It has very high-level supporters including founders of Uber and Twitter. It offers projects in different regions across different types of restoration and conservation, including blue carbon or afforestation, reforestation and revegetation.

UN Tourism runs various competitions for the best start-ups in travel and tourism, also linked to each SDG. Previous winners include Visualfy for hearing accessibility, Travaxy for accessible travel and eccocar for sustainable mobility.

TIPS

- Engaging with the travel tech and climate tech communities will create an open platform for sharing best practices on how to leverage the latest know-how to solve challenges.

- Links with academia and schools through events, workshops, open days, placements etc. will strengthen sources of innovation, awareness and change traveller behaviour for the better.

- Horizon Europe, the EU's funding body for research and innovation, has a budget of EUR95.5 billion to 2027 that can be leveraged by European businesses.

Diversification and innovation across the board

The travel industry is a complex ecosystem of different travel businesses, organizations, products and services across the supply chain, all trying to appeal to the traveller. Following decades of digitalization, commoditization has occurred in the types of travel products and services provided. On the other hand, digital has driven personalization through big data and AI. Advances in Gen AI will lead to greater levels of personalized travel services thanks to real-time and contextual data. To keep up with the eternal search for the next big thing with cachet, travel businesses and destinations are constantly looking to diversify.

So far, diversification has focused on finding new source markets to target and sell such as shifting from China to India post-pandemic. Travel suppliers are looking for literally cooler destinations such as Scandinavia or the Baltics rather than the Mediterranean. Different traveller segments are also a popular strategy – finding who will be a more resilient and higher spending source of demand. The blended traveller was a default option post-pandemic but has come under fire for contributing to overtourism due to their preference for staying in short-term rentals.

Diversification consists of many shapes and sizes. Yet the most effective are looking firstly at domestic, even day trips, to carve out resilient sources of revenue even if on the doorstep. Greater scrutiny of the environmental and social impacts is happening, for example balancing the carbon footprint of an international traveller from Asia Pacific who may stay for two weeks versus an intra-regional visitor who stays for less than 24 hours.

'It's important to take a holistic approach to tourism so that positive impacts for communities and nature are built into the design from the beginning whilst negative impacts are eliminated or reduced progressively. This also requires shaping travel from within the destinations, together with communities, rather than imposition by outbound markets. We need to help travellers to easily identify the "better choices" when choosing a travel product, and at the same time, work throughout the industry with our partners to scale responsible travel, making it the default option.'

Laura Kotyga, B2B Sustainability Manager, Evaneos

Conscious of higher prices and overcrowding in popular destinations, consumers are equally taking a more critical look at where to visit. Destination 'dupes' or alternatives have emerged as a way to enjoy similar levels of experiences but at more affordable prices. The travel industry and the media have been quick to jump on the bandwagon. However, without a well-thought-out destination management strategy that incorporates dispersal of visitors to lesser trodden places, the same problems of overtourism will continue to occur. Second- and third-tier cities may jump at the chance of being the latest place to be yet should take a long-term view.

Visitor dispersal is not a new concept and many destinations continue to reassess their trails, routes and corridors, and importantly look to spread demand through the year. Taking the pressure off peak seasons is critical to bring back control to the visitor/community experience. Mobile and geospatial data (GDPR compliant) can also be leveraged to help create new routes and spread tourism revenues and benefits where capacity and infrastructure are available.

Peru and Jordan are great examples of building new trails to take pressure off the iconic sites of Machu Picchu and Petra respectively. Even within cities themselves, visitor movements are determined by routes with greatest density of visitors congregating in the most attractive, connected and accessible places with the most activities. In Indonesia, the government announced that they would promote five new Bali's to relieve pressures on its popular destination. For some travel businesses, lesser known is a USP that appeals to their customer base. For example, Much Better Adventures promises non-touristy places and offers off-the-beaten-path treks in emerging destinations like hiking in Tunisia and Kyrgyzstan with saunas and yurts.

Diversification of traveller segments is another way to eke out value creation. Inspired by passion/interests, income, needs states, or life-stage, each visitor is looking for something more personalized and relevant to them. There is still huge scope to develop truly customized products and services that cut across the entire preferences of a visitor and their travel companions including transport, accommodation, food and drinks, experiences and other activities. Reducing the expectation/reality gap through Emotion AI will help elevate this type of customization if used ethically.

Diversification can take place in terms of how cities intra-regionally, for example, work together. With developments in urban air mobility, new hubs and corridors will open up so that there is more co-opetition between urban centres. Cities that partner closely such as Amsterdam, Helsinki and Copenhagen may see more collaboration in terms of joint promotion of visitors, so that there is a shared visitor pool that benefits all and enables destination swaps. Such partnerships have already taken place with Paris and London joint marketing together.

TIPS

- From digital twins of smart cities to digital twins of customers, data science and AI will increasingly be used to understand individuals to personalize travel products and services.

- Getting to know a customer to understand and pre-empt their different needs in different contexts will require a long-term customer strategy, incorporating customer relationship management (CRM) and loyalty.

- Creating communities of brand and destination ambassadors to tap into for shared stories and co-creation would be a useful resource of innovation.

REAL-WORLD EXAMPLE
Venice adopts a multi-disciplinary approach to innovation

Venice and its port are a pilot for the Bauhaus of the Seas Sails project (BOSS). The project funded by the EU's Horizon innovation programme aims to transform coastal communities, urban design, regenerate ecosystems and combat climate impacts. The vision is for stakeholders and communities to co-create environmental and social solutions to ensure inclusion and drive wellbeing. A regenerative menu of food and ingredients is a core element along with boosting digital inclusion for seniors.

The innovative project is multi-disciplinary embracing art, design, culture, food, technology, the environment and science. A 'drop and ripple' approach is being used to drive success in an iterative way where each location – Venice, Malmö, Genova, Lisbon, Oeiras, Hamburg, Rotterdam – shares lessons learnt that can be scaled up across the network.

Technology like blockchain and decentralized autonomous organizations (DAOs) will be used to facilitate the project. Stakeholders will form assemblies to share ideas and co-create solutions where a geospatial platform will help bring concepts and impacts to life. Adopting the latest cutting-edge technologies to enable collaboration with communities, local government and the private sector enables local voices to be heard.

TIPS

- Each destination is unique where value creation can be found in all aspects of tourism if the quality of the product matches consumer expectations.

- The time of the tourist trap is nearly at an end.

- Running small pilots helps to scale solutions quickly for a just and sustainable transformation from hyper-local to global.

- Co-creating with communities where they take centre stage in how their places are developed and promoted is a win-win for travel businesses and quality of living for local people.

- A multi-stakeholder approach to scenario planning helps to drive consensus before undertaking costly investment.

- Inter-disciplinary collaboration helps to join the dots and spark creative solutions.

Long-term vision

Near-termism remains a challenge for many democratic countries that go through political changes in government on a rolling basis. The focus on short-term goals and priorities clashes with long-term climate and biodiversity needs. Bold climate targets often are pushed back in place of more populist concerns. Taking the long view is now more essential than ever as target deadlines approach and are ultimately missed. Countries that take a long view of their strategic goals include authoritarian regimes such as China, Saudi Arabia and the UAE that equally face first-hand the impacts of the climate emergency.

Dubai (UAE) faces the wrath of climate change from flooding to extreme heat. The Dubai government is highly engaged with future-proofing its economy and building resilience through organizations such as the Dubai Future Accelerators, part of the Dubai Future Foundation. This aims to fund start-ups in areas such as digital and sustainable innovation co-creating with private and public sectors. Central to the Dubai government's long-term vision is to be a global hub for innovation and finance, attracting top talent through venture capital provided by the Dubai Future District Investment Fund. Key areas of research include inclusive and people-centric technology including AI, Gen AI, robotics and all Fourth Industrial Revolution technologies. Dubai's futures research, The Global 50, is a fascinating look into the future, outlining the most impactful megatrends and opportunities for the city to embrace. Key themes include nature, society, systems, health and transformation.

REAL-WORLD EXAMPLE
China takes the long-term view for tourism to balance rural and urban prosperity

China is a global powerhouse and major driver of global GDP. China like other countries faces challenges such as an ageing population, youth unemployment, a property crisis and climate change. It aims to see peak emissions by 2030 and reach carbon neutrality by 2060 as it transforms its economy.

Over the past decade, it has transformed into the largest spending outbound tourism market pre-pandemic thanks to a rising urban middle class. The majority of destinations around the world have targeted China outbound tourism to boost their tourism economies. It was among the last countries to remove pandemic restrictions where cross-border travel was stopped for three years, only reopening its borders in

January 2023. Its return to exploring the world especially Europe and the Americas has been slower than hoped due to geopolitical tensions. Many countries especially in South East Asia like Thailand and Singapore are opting for visa-free programmes to accelerate the return of high-spending Chinese travellers.

China has a long-term plan into the next century – to remain a global leader – which is enabled by five-year plans of which it is in its 14th Five Year Plan. China's priority is improving the quality of living for its citizens. Its economy is driven by its capital markets, innovation, technological prowess, investment, R&D, leadership in green technology like EVs and renewables as well as consumer appliances and electronics.

Its domestic tourism economy is by far the largest in the world, with billions of domestic trips taken each year, especially at peak travel times such as Chinese New Year and Golden Week (see Figure 7.1). Travel and tourism play a role in its ambitions to transform its economy to innovation and domestic consumption. It is also home to the largest global OTA – Trip.com – that is a trailblazer in AI and big data.

Revitalizing rural tourism villages is a key strategic government priority to maintain its robust domestic tourism economy and meet the SDGs. Tourism has been spotlighted to achieve China's goals for shared prosperity, where a key focus is on rural tourism to help drive wealth distribution and eradicate poverty. China has adopted a tourism+ model which Macau has also embraced which delivers an all-for-one tourism. This is where tourism is integrated with different industries such as agriculture, fisheries, e-commerce and culture for more impactful outcomes for sustainable development. Experiences are varied including home stays, culture, sports, arts, health, adventure and music.

FIGURE 7.1 Domestic tourism in China – trips and spending 2024–2029

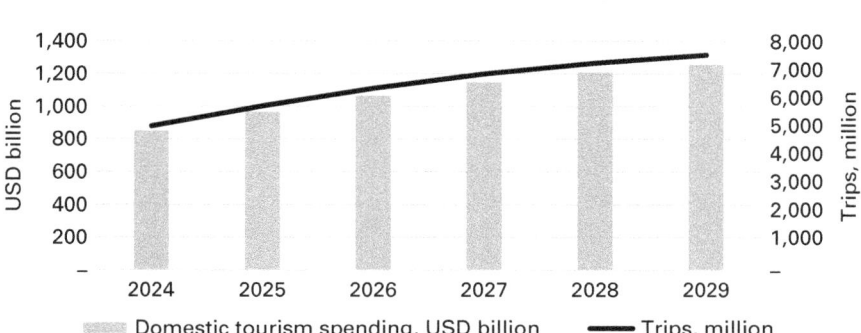

SOURCE Euromonitor International, Travel 2025 edition
NOTE Value at constant 2024 prices, fixed exchange rates

An interesting example is where e-commerce giant, Alibaba, and its online travel platform, Fliggy, worked with three villages in Hunan Province to create a digital travel guide and brand for Yongshun. The digital guide provides personalized itineraries and the ability for online bookings of hotels and tours. A livestreaming session enabled local communities to sell their agricultural produce online. Capacity building through livestreaming training for women helps to boost equality and income generation in rural areas as well as increase food security.

Its Belt and Road Initiative (BRI) speaks to China's aim to redraw the global map of influence, revitalizing the ancient Silk Roads through investment in trade, infrastructure, transport, energy, digital and tourism. The BRI network expands even to Latin America. BRI celebrated 10 years and its legacy will continue to grow and evolve.

The Greater Bay Area is a megalopolis of nine Chinese cities including Guangzhou and Shenzhen with the special administrative regions of Hong Kong and Macau. An impressive multi-modal integrated transport and infrastructure system has been created including high-speed rail and the Hong Kong-Zuhai-Macau bridge, the longest sea crossing bridge in the world.

TIPS

- China's approach offers many lessons on how to build resilience and sustainability through balancing rural and urban development.

- Taking a long-term view of how to build strong foundations for domestic tourism is central to transformation.

- Focusing inwardly helps to deliver higher quality, truly authentic travel experiences and products that resonate with the domestic market and will be more appealing to international markets.

- With food insecurity increasing in line with population growth and carbon emissions from global warming, connecting tourism and agriculture is vital.

- Combining agritourism with e-commerce opens up new revenue streams especially for youth and women for equality.

Tourism futures disservice

There are certain future pathways that may arise if the travel industry continues to grow unchecked where the cost-benefits are tipped against local communities and destinations. Already there are examples of destinations and communities refusing to promote themselves, remaining behind closed doors as gated communities. Some may even go as far as to opt out of tourism, turning their back on visitor revenues and the jobs created to find alternatives. Such 'tourism exclusion zones' or 'tourist-free zones' may set a precedent especially in the Mediterranean where carrying capacity has been breached in fragile areas.

Enclave tourism, where visitors stay in gated resorts and do not engage with the local community, will take on a new perspective as the enclave will be demarcated by communities where they will or will not welcome visitors. This will be a change from the traditional enclave tourism where resorts are all-inclusive so that visitors are separate from communities, often presented as due to crime but mainly to ensure captive spending. One of the primary reasons to travel is to engage with different cultures and people so creating zones would run counter to the benefits of welcoming visitors and cultural exchange.

Consumers are constantly looking for new and undiscovered destinations. Yet any 'secret' does not remain hidden for long. Where there is exclusive access, it leads to financial exclusion where only the higher income brackets will have the means to access the gated communities in the same way as they have to private members' clubs.

Traveller segmentation is a common practice for travel businesses and destinations. Already there are signs that climate impacts alongside the standard length of stay and propensity to spend are being factored into which future travellers to target. Taking this to the extreme, for example social profiling such as cross-referencing social media for anti-environmental or anti-social behaviours and attitudes would be unethical.

New international routes are often rolled out by airports and airlines without the input from destination partners which again sets up for potential problems and clashes if not managed in a sustainable and fair way. Multi-stakeholder collaboration is essential especially across sectors like energy, transport, food and drinks, retailing and hospitality for the transformation required.

If the travel industry fails to decarbonize in the way set out in the Paris Agreement, and it takes an increasingly larger share of the carbon budget, a worst case scenario could occur where travel becomes the new smoking. This would lead to punitive regulation as seen in the past few decades with the tobacco industry. Court cases could be brought by communities saying that travel and tourism are bad for their health and wellbeing due to the carbon footprint of visitors. This situation could take place especially in places with high levels of pollution where the air quality is hazardous to health. This extreme scenario must be avoided at all costs as it would not benefit anyone in the long run.

TIPS

- Exclusivity will always play a role for some yet having an open approach to visitation helps to democratize travel experiences and ensure inclusion.

- Educating visitors about how to be a sustainable and conscious traveller whether at the time of booking, marketing campaigns or taking a pledge are excellent ways to set expectations and pave the way for mutual respect.

- It comes down to joined-up planning and destination management to ensure that closed-off pathways and zero tourism communities do not transpire.

- Airlines need to get round the table with their destination and community partners and not make route decisions in silo with airports alone.

- There's an opportunity to ask the community to say which airline route they would like to see introduced for inbound/outbound shared benefits.

- Decarbonizing at speed and scale will ensure that travel and tourism do not become the new smoking.

Shared future for all

The benefits of travel and tourism are well documented in terms of boosting intercultural exchange, promoting diversity and inclusion and a sense of wellbeing and transformation for many. Adopting a regenerative approach works externally as well as internally, bringing a deeper understanding and awareness of local communities and cultures. Travel is a force not just for good but personal and collective benefits. It requires that it is managed in a measured, sustainable and inclusive way. As prices continue whether through taxation or fees to fund the sustainable transformation, it is important that income does not stand in the way of access to travel experiences. Flexible pricing and payment innovations will be required, but equally important, different ways of enabling travel to remain open, such as funding for youth, education and upskilling, for example.

Depending on the level of extremity that AI will have on jobs, one financial lever could be universal basic income (UBI) linked to accredited sustainable travel providers and climate-positive destinations. UBI would first and foremost be used to alleviate poverty through a guaranteed income. With more scientific proof about the health and wellbeing benefits of travel and tourism, especially for children, there could be a case to incorporate travel credits in UBI. There is already much research on how wellness tourism in particular has clear positive outcomes on physical and mental health and holistic wellbeing. Tourism has been shown to enhance quality of living and boost happiness. Physical activities such as nature-based and adventure travel release feel-good endorphins and are mood-boosting. There are multiple psychological benefits from socializing, meeting new people, discovering new places, boosting empathy and under-standing.

As seen after the pandemic, tourism vouchers and free visitor experiences were offered in countries like Japan to boost economic recovery and social wellbeing. Any link to UBI would require vetting of travel businesses engaged with the SDGs and following science-based targets.

With the so-called Godfather of AI, Professor Geoffrey Hinton, warning about the need for UBI due to job losses caused by AI and automation, the case for UBI is gaining traction including pilot schemes in the UK.

TIPS

- The holistic benefits of travel need to be more grounded in neuroscience to ensure that travel remains open to all.

- Travel and tourism have a role to play in universal income creating an opportunity for travel businesses to contribute to the collective wellbeing.

- It is important to avoid polarizing the case for travel – whether a total ban or a free for all.

'In the future, travel businesses will need a moral licence to operate and deliver net positive impacts.'

Vicky Smith, Founder, Earth-Changers

REAL-WORLD EXAMPLE

Brazil and the urgent need to restore the Amazon and Pantanal through regenerative tourism

Brazil is a vast country with several biomes. The plight of the Amazon is global news as one of the world's largest natural ecosystems and richest sources of biodiversity.

The rate of deforestation slowed in 2023 after the change in government, yet stood at 17 per cent where anything over 20 per cent would breach a tipping point.[15] This would lead to the Amazon no longer able to restore itself and would turn it into dry grassland and savanna which would release more carbon into the atmosphere. Areas in the Amazon under Indigenous People's control act as carbon sinks while other areas that are not under Indigenous control are sources of carbon. Deforestation, agriculture and logging are some of the main reasons for the Amazon's degradation.

Cerrado is home to 5 per cent of the world's wildlife and plants but does not enjoy the same fame as the Amazon.[16] Despite success in the Amazon in bending the curve, the Cerrado biome has experienced a loss of half its natural state and is degrading three times faster than the Amazon.[17] Regenerative models such as EcoAraguaia and its Farm of the Future, work with Ecosystem Restoration Communities, and advocate a regenerative approach to farming combined with tourism. At the crossroads of the Amazon, Pantanal and Cerrado, the visitor experience features engagement with farming, community, the environment and nature – co-created with the local community. This helps to build a vibrant rural community and reverse the urbanization trend, providing opportunities for young people to stay.

In Mato Grosso do Sul, destinations like Bonito fly the flag for regenerative tourism. Bonito has signed the Glasgow Declaration and works with the Green Initiative that accredited it with meeting its climate neutral targets. Carbon measurement is fundamental to its climate action plans where each visitor has a carbon footprint of 0.73 tonnes.[18] Core focus areas are decarbonizing transport by moving to electric mobility solutions and solid waste management. Bonito aims to be a role model for climate smart tourism acting as a cluster of innovation across the private and public sectors. Ensuring the right funding is in place to accelerate the necessary transition to net zero is vital, where accreditation helps in that process.

In Bonito, the ecotourism resort, Estância Mimosa, achieved carbon positive status through the Green Initiative, and has shown to extract more carbon than it produces across scope 1, 2 and 3 emissions. This climate-positive success was helped by its forest preservation and regeneration and its carbon capture capabilities where 84 per cent of the property is made up of native forest.[19] The eco-resort offers nature-based adventure experiences such as the floating tour in the Rio da Prata with pools and waterfalls. Even the boats run on solar power. The emphasis is on local produce. The business model is highly successful with multiple awards, including TripAdvisor Traveller's Choice where regenerative and climate-positive experiences elevate the joy of engaging in nature for visitors.

TIPS

- Climate change, global warming, food insecurity and job losses due to AI are creating a perfect storm. Travel businesses need to adapt and embrace new regenerative, climate-positive business models and create transformative experiences for visitors.

- Indigenous and local communities are the most effective guardians of life on Earth and travel businesses must enable local knowledge to radiate and inspire others.

- The beauty of agri-tourism is that it supports food security, jobs and income for communities while providing local and authentic gastro experiences for visitors which is a win-win.

Measuring success

The travel industry is highly fragmented yet needs to come together to move along the same pathways to sustainable transformation. Aviation tends to operate in silo from tourism while travel tech at times seems elusive from MSMEs. There are start-up communities that intersect tackling climate, inclusion and travel tech. Silos must be removed if travel businesses are to have a chance of a successful future.

REAL-WORLD EXAMPLE
Machu Picchu – world's first carbon neutral destination through the power of partnership

Machu Picchu is a wonder of the world and a UNESCO World Heritage Site as a sacred ancient city attracting millions of visitors each year. The associated downsides as an iconic destination led to restrictions on visitor numbers, allocated time slots and advance booking with 4,200 Inca Trail tickets available per day. There are official tour companies and a limit of 10 people per guide, with designated circuits to be followed.

Machu Picchu is the world's first destination to be awarded Carbon Neutral status by Green Initiative in 2021 and again in 2024. It achieved this by meeting its targets on its path to reduce emissions by 45 per cent by 2030 and reach net zero emissions by 2050. Green Initiative is an advisory and certification organization, approved by UN Tourism and OnePlanet, helping to meet the Paris Agreement targets in line with the Glasgow Declaration. Machu Pichu's success in decarbonization is predicated on the strength of the public-private partnership and strategic alliance including PromPeru (national tourism board), Inkaterra (award-winning luxury hotel company), the local municipality, AJE and TetraPak among others. Inkaterra is equally impressive as the world's first carbon positive hotel, as accredited by Green Initiative, promoting regenerative tourism.

Carbon neutrality has been achieved through various measures by going circular including waste and plastics management, restoration and planting native trees for regeneration and resilience. For example, vegetable oil was converted to biodiesel. Everyone in the whole community was encouraged to contribute and reduce the local carbon footprint. Compared to the 2019 baseline of 7,144 tCO2e, there was a decrease of 19 per cent in terms of carbon emissions in 2024.[20]

Alternative hiking trails were also developed to relieve further pressures on Machu Picchu and disperse tourism spending, by visiting local villages. Peru aims to decouple tourism growth from carbon emissions by taking a three-pronged approach to decarbonization: destinations, businesses and visitors.

Summary

The framework to achieve the 17 UN SDGs by 2030 and reach net zero emissions by 2050, if not before, is firmly in place. The pathways are laid out clearly for governments and businesses, while consumer societies understand what is required – sustainable transformation. Yet progress is not linear as national commitments do not go far enough and a business-as-usual mantra persists. A value growth mindset, decoupled from carbon emissions is required, which represents a shift away from the volume growth mindset. Some at the cutting edge of climate change and biodiversity loss are already in the fight for survival, while those with foresight such as adventure and transformative travel embrace regenerative solutions. This change in approach requires a long-term view with the tactical know-how and tools to adapt in the short term.

All travel businesses and destinations potentially hold the key to unlock transformation to deliver the benefits that tourism can provide if managed in a measured and inclusive way. Leveraging tech for good allows faster and smarter decision-making. Tech sparks creative thinking, agility and innovative solutions. For example, real-time geospatial data from sensors feeds models and digital twins to drive resilience and future-proof businesses. The role of tech is as an enabler of solutions and collaboration, empowering local communities to have a louder voice, contribution and governance. The combination of tech, nature, climate and travel is a powerful weapon to ensure a sustainable and just transition that leaves no one behind.

The stakes are high. There are signs of what may transpire for travel and tourism if not transformed into a value growth paradigm. The repercussions would multiply from institutional anti-tourism, community backlashes and tourist-free zones, to punitive taxation on businesses and consumers for starters. Travel and tourism should never be seen as harmful to people and nature as they have the power to restore. Taking positive steps to create a climate action plan now is the optimum way to tackle the challenges and maximize the opportunities that lie ahead.

Endnotes

1 European Union Net Zero Cities (2024) netzerocities.eu (archived at https://perma.cc/R6ML-4KZZ)

2 Ibid

3 Ibid

4 Circularity Gap Report (2024) lookerstudio.google.com/reporting/3ae548bd-15b8-4bc8-8ff1-9bd254bf6cc9/page/p_220hq5p23c (archived at https://perma.cc/2FN2-SVG2)

5 Carbon Neutral Cities Alliance (2024) carbonneutralcities.org/cities/copenhagen/ (archived at https://perma.cc/66D3-VM97)

6 Catalan News (2024) www.catalannews.com/business/item/tourists-in-barcelona-spent-96-billion-in-2023-up-147-from-2019 (archived at https://perma.cc/2NXF-42R3)

7 EuroCities (2024) eurocities.eu/stories/barcelona-shapes-the-future-of-city-planning/ (archived at https://perma.cc/CS3S-V9AT)

8 EU Joint Research Centre (2024) joint-research-centre.ec.europa.eu/jrc-news-and-updates/global-warming-reshuffle-europes-tourism-demand-particularly-coastal-areas-2023-07-28_en (archived at https://perma.cc/WL9G-PRL5)

9 Belize Tourism Board (2024) www.belizetourismboard.org/wp-content/uploads/2024/02/BTB_NSTMP_Update_Adapt_Plan_Final_Master_n-compressed.pdf (archived at https://perma.cc/8CXL-2QFZ)

10 Our World in Data (2024) ourworldindata.org/un-population-2024-revision (archived at https://perma.cc/RS9B-XYPR)

11 United Nations (2024) unstats.un.org/sdgs/report/2023/goal-11/#:~:text=The%20world's%20population%20reached%208,70%20per%20cent%20by%202050 (archived at https://perma.cc/FKZ2-7YFX).

12 World Economic Forum (2024) www.weforum.org/publications/travel-tourism-development-index-2024/interactive-data-and-economy-profiles-afaa00a59c/ (archived at https://perma.cc/T83C-XKDN)

13 European Union Aviation Safety Agency (2024) www.easa.europa.eu/en/light/topics/fit-55-and-refueleu-aviation (archived at https://perma.cc/B7YK-3WLX)

14 Civil Aviation Authority of Singapore (2024) www.caas.gov.sg/docs/default-source/docs---so/singapore-sustainable-air-hub-blueprint.pdf (archived at https://perma.cc/QQQ8-HYVT)

15 Amazon Conservation (2024) www.amazonconservation.org/the-challenge/threats/ (archived at https://perma.cc/XMP2-44FA)

16 WWF (2024) www.worldwildlife.org/ (archived at https://perma.cc/EPE7-R2HJ)

17 Mongabay.com (2023) news.mongabay.com/2023/08/a-tale-of-two-biomes-as-deforestation-surges-in-cerrado-but-wanes-in-amazon/ (archived at https://perma.cc/3SGG-9HSA)

18 One Planet Network (2023) www.oneplanetnetwork.org/news-and-events/news/interview-bruno-wendling-director-president-mato-grosso-do-sul-tourism (archived at https://perma.cc/E44E-GVGH)

19 Visit Mato Grosso do Sul (2023) www.visitms.com.br/en/noticias/sustentabilidade/atrativo-em-bonito-e-o-primeiro-do-mundo-a-receber-a-certificacao-climate-positive/ (archived at https://perma.cc/HJB3-76XJ)

20 Green Initiative (2024) greeninitiative.eco/2024/06/29/from-heritage-to-habitats-the-journey-of-sustainable-conservation-from-machu-picchu-to-the-amazon/ (archived at https://perma.cc/GU9J-D7ZK)

Conclusion

No more business as usual

Existential threats require a pragmatic approach

Undoubtedly, the travel industry is an important economic powerhouse accounting for 9.1 per cent of global GDP, creator of 330 million jobs and often hailed as a force for good.[1] On the flipside, it faces a major paradox due to its large carbon footprint, amounting to 8 per cent of global carbon emissions where transport is the biggest culprit.[2]

Travel and tourism are at the forefront of change, constantly facing new threats, challenges and opportunities. There is a myriad of drivers to contend with: including geopolitics, wars, uprisings, terrorism, pandemics, health emergencies, macroeconomics, consumer demographics, socio-cultural shifts, climate impacts, natural disasters, megatrends and technology. Disruption is never far off where change as constant stands the test of time. Despite the latest technology, forecasting tools and advanced scientific knowledge to deliver the SDG agenda, there are many unknowns. However, what history tells us is the need to be prepared and embrace change with a growth mindset.

Travel businesses temporarily faced a worst-case scenario with zero discretionary travel at the height of the pandemic. This global shutdown is a harbinger of what could befall the sector if it does not sort its act out and transform to a value-driven model, infused with purpose and passion. The next time global travel bans come into force, these may be enforced by governments due to climate, not health reasons. The fact that the carbon budget will run out by 2030 in the global warming scenario of 1.5°C, if not sooner, speaks to the urgency of the need for transformation across all sectors and societies.[3] This is an opportune moment for travel to lead.

The existential threat has not gone away with the return of peak visitation, and dawn of a new golden age of international travel, smashing all previous records. At the heart of travel lies an uncomfortable truth: travel is a force for good and a cause of harm if not managed correctly. The pandemic has not solved this duality as business as usual has returned. The majority of travel businesses and destinations are enjoying the last hurrah, continuing to travel with a significant carbon footprint, with little regard for long-term impacts. The forecasts point to an onwards and upwards linear trajectory for demand and supply, perpetuated by rising middle-class incomes in emerging markets and Western consumers unwilling to alter their travel habits.

The new growth era exacerbates the same-old problems of overtourism, inequalities, anti-tourism sentiment and counter-tourism measures such as taxation and visitor limits. There is a place for these interventions, but prohibitive measures can be avoided if the entire system transforms at scale and rebalances to put local people, communities and places first. It's not just tweaks around the edges, but a full scale reset so that there is mutual benefit, trust and shared value built into a symbiotic host community-visitor relationship.

Global frameworks are in place to guide the 2030 and 2050 agendas yet must not be rigid and should be open to new research, science, citizen science and local empirical knowledge. Empowering local communities, businesses, MSMEs, grassroots organizations and non-profits to have an equal voice and role to play in their own future development is fundamental. Without the micro view, the weaknesses in the global system will continue to hamper true progress in sustainable and regenerative tourism development in the places that desperately need them.

No one business, destination, vertical or industry can work alone. Partnership and collaboration across the entire chain is required to break silos, remove complexity and the challenges of being such a highly fragmented ecosystem where digital tools can play an important role.

Every year new record temperatures are broken and tipping points for climate and biodiversity become ever more real. Resistance, denial or passing the buck from government to business, business to consumer, and back again, is non-sensical and creating a stalemate. The time for action and engagement in the triple planetary emergency of climate, pollution and biodiversity is now. It is not just an uncomfortable truth; it is a scary truth. This requires a positive and pragmatic approach to ensure that severe scenarios become less severe and more manageable as the 1.5°C global warming scenario proposes.

Bold steps forward for positive action

Tinkering around the edges is not an option, instead it requires adopting bold climate action, starting small and scaling up. Even the most top-down organizations like the UN are calling for a paradigm shift to ensure human-kind can thrive. The tools, resources and means are all available so failure to act is like signing a death warrant for future generations.

We remain in a code red for humanity. The frameworks to achieve the 2030 Global Goals and reach net zero emissions by 2050 are in place to guide the transition. The business case for adopting innovative sustainable solutions is ever more pressing as the target milestones of 2030 approach and action is accelerated to reach net zero emissions by 2050. Destination best practices need to continue at rapid speed, such as spreading demand more equitably geographically and throughout the year, boosting average length of stay along with revenue kept locally. Diversification of source markets more aligned in terms of spend and positive travel behaviours is desired, while creating a new playbook, mixing up old and new destinations and going off the beaten path for shared prosperity.

Nevertheless, tensions in popular destinations are spilling over where local communities are rising up to say enough is enough, taking action into their own hands. The benefits of travel and tourism of cultural exchange, empathy, tolerance and understanding are stifled when destinations are mismanaged and inequities spiral. Government levers such as visa programmes, taxation and visitor limits can only patch things up to a certain extent.

Transformation of any form is hard and takes time. Pressures and tensions are difficult to navigate and come from multiple different sources: legisla-tors, investors, employees, communities and consumers. Greenwashing versus greenhushing is a major challenge: you are damned if you do and damned if you don't. Consumers find it hard to constantly do the right thing by the environment with green fatigue and malaise setting in. Younger generations feel the force of climate injustice and resent the state of their inherited planet. People have complex needs: they want instant gratification, and a constant stream of new and unique experiences tailored to them, all delivered guilt-free. Sustainable innovation is therefore the key to ensuring that travel and tourism remain open to future generations.

Adopting an innovative mindset looking across the travel ecosystem from scopes 1 and 2 and the more challenging scope 3 emissions will help identify where to lean in and focus on decarbonization where travel and tourism are a major contributor. There is a huge opportunity for travel businesses to spearhead incentivizing sustainable traveller behaviours, empowering communities, employees and customers to be the agents of change. With trust being the most elusive commodity to earn, transparency and technology can play a role in building trust into the travel ecosystem. Moving to accreditation is a great way to deliver trust. The best places to start are SBTi, B Corp and GSTC along with signing the Glasgow Declaration to make a formal commitment to climate action.

Fresh face of the future global traveller

One of the most exciting future outcomes is welcoming a whole new generation of diverse travellers from emerging markets to the world of travel. Billions of young, middle-class consumers will start exploring their home countries and eventually travel further afield, exploring intra-regionally and internationally. The future traveller will be diverse, digital and driven by their passions, interests, curiosity, wellbeing and aspirations. Understanding new trends, traveller segments and influences will be a constant undertaking to keep ahead and align values from community to visitor.

Gen Alpha, a generation of keybox kids, will demand inclusivity, hyperconnectivity, creative expression and authenticity. Already there is a backlash against industrial growth and globalization in the pursuit of profit at all costs. The Baby Boomers are ageing but not giving up their dreams of travel, requiring an inclusive approach to ensure no one is excluded.

As transport connectivity continues to grow, it is hypocritical to expect future travellers from large populous emerging regions like Asia Pacific and Africa to not explore the world. This requires ensuring a just transformation to a sustainable travel ecosystem. Accounting for future migration flows will also help to ensure that pressures on infrastructure and resources are not stretched, spilling over into conflict.

As travel matures, it will premiumize and polarize. Green charges and taxation are being introduced to fund the transition, where higher prices will increasingly be prohibitive. Rising prices will inevitably force out many in the lower- and middle-income brackets. It is, however, vital that travel and tourism remain open to all. This may require a deeper dive into the wellbeing benefits of tourism which may lead to government interventions to ensure it remains available especially to those who most need it.

Every single travel business and destination, large and small, needs not just to be primed for sustainable development but on the pathway whatever stage in the journey. An honest conversation needs to happen especially with large travel businesses where it will be necessary to retire unsustainable products and services that cause the most negative impacts. Mass market commoditized products and services should be the starting point, namely all-inclusive package holidays and large cruise ships. Other big question marks hang over long-haul flights, private jets, frequent flier programmes as well as stricter regulation to enable affordable housing and fair wages. Commoditized, identikit and carbon intensive are not the attributes that people look for when choosing a holiday, they seek authentic experiences that align with their values. Staycations offer a low carbon alternative and could do with a makeover to make them more appealing.

Technology as an enabler of sustainable travel

The success of travel and tourism is wrapped up with technology, woven into the entire customer journey and across the supply chain to provide a seamless travel experience. With the advent of each new breakthrough and exponential leap forward, travel businesses jump to integrate it into their systems and services as seen with Gen AI. Technology for technology's sake is often an error made, as businesses try to keep up with the competition. Gen AI is ushering in a new era of personalization, working off ever more data but unleashing a new host of challenges from data integrity, trust and ethical concerns where regulators are struggling to keep up.

With the next generation of internet, consumers will be hyper-connected and even more demanding. If technology runs in silo and is not fully integrated and aligned with delivering positive impacts for communities and places, then it will hinder rather than help. Technology is not a silver bullet, but it is a key enabler of the triple planetary agenda if led by purpose. Providing up-to-date and timely if not real-time data on environmental, biodiversity and social impacts can be ramped up exponentially. Fundamentally, the move to ubiquitous digitalization demands gold standards in data privacy and security.

Online community platforms, resident surveys and apps that measure citizen happiness such as the Happiness Meter in Dubai will be increasingly useful to ensure that communities are engaged with how travel and tourism are developed locally. Giving local people a voice and platform for engagement and participation will help pre-empt future challenges to urban planning where tourism plays a key role.

As digital hyper-connectivity increases so will social challenges such as isolation and loneliness where travel can play an important role in transformative and restorative experiences. Nature-based experiences and engaging with communities, biodiversity and the natural world are an essential antidote to the negative effects of ever more digitalization. Taking a dual approach to looking internally and externally in terms of what local people and places need will help create more positive outcomes.

Travel – central pillar of system-wide transformation

Travel is at a critical juncture. It can continue to change piecemeal, or it can embrace system-wide transformation beyond its boundaries to meet the desired outcomes of the climate targets. Collectively the current government pledges on NDCs do not go far enough to deliver success. Travel's dependency on transport, energy, construction and infrastructure puts it at the frontline of change but governments tend to be faint-hearted and transient. It is up to travel businesses of all sizes to pick up the baton and run with it. There are pockets of great work being done, but it is highly fragmented and not yet at scale from grassroots upwards. Europe is experimenting with new innovative and creative approaches. Mandatory sustainability reporting like the Corporate Sustainability Reporting Directive (CSRD) helps to mainstream and improve ESG data alignment. It will be necessary to tackle some of the more opaque financial areas of the travel industry such as making asset holders, venture capital and real estate investment trusts (REITs) more accountable for their actions and scopes 1, 2 and 3 emissions on the ground.

Technology can power operational efficiencies to dramatically reduce carbon emissions. Decarbonizing transport is a worthy cause and shifting to a more effective multi-modal transport system helps solve the last-mile challenges. Aviation as the hardest sector to abate needs to be tackled head-on by regulators. Travel businesses with flight-free options show that the 'all you can fly' flight subscription from Wizz Air is not the right path for innovation. Creative solutions are found at the intersections of tourism, nature, climate and technology. Throwing purpose into the mix makes transformation resilient, adaptable and inclusive. Taking a leaf out of start-ups, adopting a test and learn approach, ensures agility and the ability to change course without delay.

Although Gen AI is only at the beginning of its journey, future generations of travel businesses and entrepreneurs will be AI native. With a new age of quantification thanks to hyper-connectivity and the Internet of Everything, there will be so much data that it will be overwhelming. This requires upskilling of employees to embrace data and metrics into their day to day.

With a system-wide approach, the long-term and short-term effects of travel such as demand forecasting or use of agriculture for SAF can be addressed thanks to predictive analytics and digital twins of the smallest of destinations to mega cities. Taking a systems approach will help to identify and balance the trade-offs and benefits that different actions could entail to ensure the optimum outcomes.

Done in a holistic way, travel and tourism can contribute to a virtuous circle of positive impacts for local communities and places. A regenerative approach, that embraces nature and biodiversity – even in the most urban areas – will put travel businesses in the best position for sustainable value creation. The future should enable thriving not just surviving.

Pathways to enduring success

No resilience without purpose

Travel businesses on the ground are in a unique position as stewards of their local people and places. Resilience to adapt requires taking a more agile view, working towards long-term goals with tactical steps along the way. Threats are ever-present and travel has proved to be resilient, adapting to the pandemic but it has not learnt to transform to a sustainable model driven by value and purpose. Following and acting according to the science, such as going B Corp, joining the SBTi and signing the Glasgow Declaration are first steps towards regenerative and net positive effects. It is important to create positive feedback loops at destination level, adapting and mitigating climate impacts, reversing deforestation and biodiversity loss while preserving cultural diversity.

The complexity of the challenges requires collaboration across the board, taking account of all stakeholders, and taking interdisciplinary solutions to a new level of transdisciplinary approaches. Nothing is off the table in a co-creative process, inspiration and innovation can come from any source. Travel sits at the intersection of a multitude of different sectors and can steer and guide how to bring in positive experiences for visitors and locals. Some of the most interesting examples take inspiration from ancient wisdom and sit at the intersection of regenerative practices such as Indigenous agri-tourism and food tourism like Regenerative Vanua (Vanuatu).

The macro-economic case for climate-positive action has been outlined clearly. The cost of inaction and damage mitigation will be six times higher by 2050 than the cost of maintaining global warming at 2°C and reduce income per capita.[4] Dealing with climate risks and damages from temperature increases and precipitation will be a hugely expensive business over the decades to come, one that will increasingly be factored into public and private finance and insurance. Climate-positive action for travel businesses is not just common sense, it makes financial sense with a multiplier effect.

Sustainable transformation within reach

There are constant shifting sands. It is therefore important to take a step back from the noise coming from both sides of the extremes. In the same way that economic growth is being decoupled from climate and carbon emissions, there also needs to be a similar decoupling of travel and tourism from political soft power.

Humanity has never had access to so much information, technology and tools to help deliver its goals. Yet, every year, Earth overshoot day comes earlier as the transition to the circular economy is not fast enough. A change in mindsets, narratives and storytelling is fundamental to elicit the required behaviour changes. Nudge theory from behavioural science is not new but is still a useful way of incentivizing better travel habits and choices. The nudges just need to be more visible, easy and trustworthy.

Some will call for a total stop to travel and promote degrowth. Even those at the cutting edge of progressive destination management like Amsterdam admit this is not a viable option. In certain places, communities will resist or even reject visitation. There will always be growth in top-tier destinations. It needs to be smarter growth that creates shared value where values align with communities and their aspirations.

Instead of Gen AI being a distraction with its promise of the ultimate personalization, it is another tool to be harnessed for destination management and power digital twins to optimize and future-proof cultural and natural assets, infrastructure, cities and sites.

Importance of storytelling to change the narrative

With overtourism back in the global headlines, there is a danger that travel and everything that it represents – human connection and cultural understanding – will be stigmatized. Travel cannot risk becoming the 'new smoking' in terms of being bad for the planet and people's health and quality of living.

This is challenging especially where countries rely so heavily on tourism, often as there is no alternative. A balance needs to be struck and action accelerated to transform to a net positive model. Giving back more than extracting charts the best pathway to sustainable success. If airports can be net positive hubs of energy by 2050, so can many hotels, campsites, short-term rentals, attractions and transport hubs.

When it comes to changing the narrative, the way that travel businesses, destinations, communities and especially the media talk about travel must change. This entails a well-overdue review of the lexicon and taxonomy. For example, terms such as 'tourist', 'traveller' and 'tourism' increasingly have negative connotations especially in communities where overtourism is rife. Even the term 'industry' has extractive and exploitative connotations where 'ecosystem' would be a preferable term. It makes perfect sense for a sector that is so dependent on the planet to take a leaf out of nature's book, adopting regenerative and nature-based terms and solutions.

Travel is also about people, their stories, histories and the billions of diverse voices. Some voices speak louder than others. Identifying those voices on the ground that have the sharpest insight into impacts should be a priority. Passionate and engaged micro influencers are everywhere, such as the barista in the local independent coffee shop or a campsite owner who grows their own produce. Spotlighting unique insights and original stories told in an unfiltered way can speak volumes over the noise of macro influencers, celebrities and a national tourism board's social media accounts (see Figure C.1).

FIGURE C.1 Sustainable pathways for a value paradigm in travel

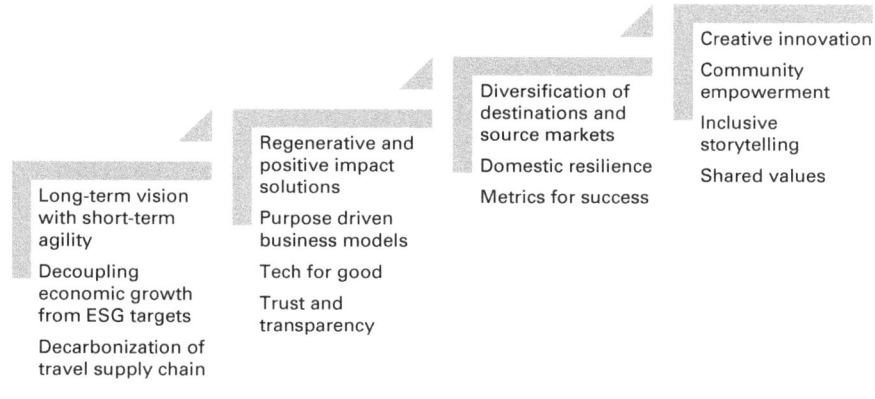

Long-term vision with short-term agility

Decoupling economic growth from ESG targets

Decarbonization of travel supply chain

Regenerative and positive impact solutions

Purpose driven business models

Tech for good

Trust and transparency

Diversification of destinations and source markets

Domestic resilience

Metrics for success

Creative innovation

Community empowerment

Inclusive storytelling

Shared values

SOURCE Caroline Bremner, author

Future-proofed

With a new era of automation unleashed by Gen AI and Industry 4.0, reskilling the workforce will take on ever greater importance. Travel employees of the future will be older, digital natives yet face the need to upskill and reskill with greater frequency. As AI starts to lead to greater job losses and inequalities, talk of Universal Basic Income will amplify. Data will be the common denominator, where measurable impacts will be standard, led by climate science. A new era of impassioned, empowered travel climate activists is on the cards.

Starting local and from early years, travel businesses can take the future vision of tourism as a pathway that aligns with passion and purpose into primary schools. Creating local climate ambassadors of every young person would be a fine legacy for the planet that travel should aim for. Travel businesses can empower their local communities, promoting nature-based and net positive solutions as part of their stewardship. Optional climate service could be created to ensure greater linkages between businesses and communities where everyone has a role to play.

There is no written guarantee that the travel industry will succeed in the future even if it transforms, but being ready and willing to adapt helps to ensure more positive outcomes. Travel alone is not a saviour but an enabler of positive change. Future-proofing travel is a means to create a world fit for purpose that the next generations will be proud to explore, experience and care for. Based on shared values, stories and purpose, travellers of today can take an active role in creating a resilient and sustainable travel industry for the future travellers of tomorrow.

Endnotes

1 World Travel and Tourism Council (2024) wttc.org/ (archived at https://perma.cc/38DS-23JQ)

2 Ibid

3 CarbonIndependent.org (2023) www.carbonindependent.org/ (archived at https://perma.cc/CW2H-58ZT)

4 The Guardian (2024) www.theguardian.com/environment/2024/apr/17/climate-crisis-average-world-incomes-to-drop-by-nearly-a-fifth-by-2050 (archived at https://perma.cc/RVC5-VZ78) from www.nature.com/articles/s41586-024-07219-0 (archived at https://perma.cc/UN3D-D9EK)

ADDITIONAL RESOURCES

Organization	Resources	Links
Adventure Travel Trade Association (ATTA)	Adventure travel association with useful research	www.adventuretravel.biz/
B Corp	ESG impact certification	www.bcorporation.net/en-us/certification
Circle Economy	Digital tool for circular scans of cities	www.circle-economy.com/digital-offering
Climate Central	Coastal Risk Screening Tool	coastal.climatecentral.org
	Climate forecasting	www.climatecentral.org/climate-matters/picturing-our-future-CM
Copernicus	Climate change, weather	climate.copernicus.eu
Future of Tourism Coalition	Overarching non-governmental organization with guidelines on sustainable tourism	www.futureoftourism.org
Ganbatte	Data on circularity	ganbatte.world/cities
Glasgow Declaration on Tourism	One Planet Network	www.oneplanetnetwork.org/sites/default/files/2022-02/GlasgowDeclaration_EN_0.pdf
Global Sustainable Tourism Council	International body for sustainable tourism standards and accreditation	www.gstcouncil.org
Google	Travel Impact Model	travelimpactmodel.org
Greenview	Sustainable lodging benchmarking tool	https://greenview.sg/ https://greenview.sg/services/chsb-index/

(continued)

(Continued)

Organization	Resources	Links
Horizon Europe	EU funding for innovation and research	https://ec.europa.eu/info/funding-tenders/opportunities/portal/screen/home
International Energy Agency	Transport and emissions	www.iea.org
Intrepid Travel	Open-source carbon label guide	www.intrepidtravel.com/uk/download-carbon-label-guide
Nature Positive Organization	Nature positive resources	www.naturepositive.org
Pacific Asia Travel Association	Members association for Asia Pacific	www.pata.org
SBTi	Collaboration for net zero targets	sciencebasedtargets.org
Sustainable Travel International	Sustainable tourism strategies and carbon offsetting for businesses, destinations and individuals	https://sustainabletravel.org/
Taskforce on Nature-related Financial Disclosures	Incorporating nature into business decisions; useful resources on how to get started	tnfd.global
Tourism Panel for Climate Change	Insightful research and action on climate change transformation in tourism	tpcc.info
The Travel Foundation	Non-profit with useful research including Invisible Burden and Envision 2030	www.thetravelfoundation.org.uk
UN Global Compact	Voluntary responsible business scheme	unglobalcompact.org
UN Tourism	Global tourism leadership, research and advocacy; includes travel and tourism dashboard	www.unwto.org
World Travel and Tourism Council	Global association for travel businesses with an extensive research hub including economic impact, social and environmental data	wttc.org

INDEX

NB: page numbers in *italic* indicate figures or tables

Looking for another book?

Explore our award-winning
books from global business
experts in Marketing and Sales

Scan the code to browse

www.koganpage.com/marketing

More from Kogan Page

ISBN: 9781398609501

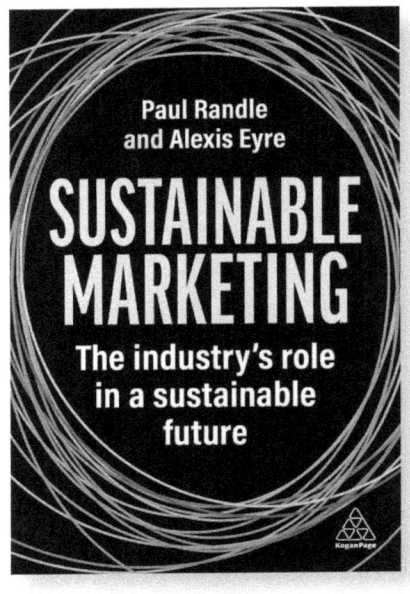

ISBN: 9781398613133

ISBN: 9781398603349

ISBN: 9781398604049

www.koganpage.com

www.ingramcontent.com/pod-product-compliance
Lightning Source LLC
Chambersburg PA
CBHW042157030325
22907CB00049B/1603